BUILDING

FOR

EVERYONE

BUILDING
FOR
EVERYONE

EXPAND YOUR MARKET
WITH DESIGN PRACTICES FROM
GOOGLE'S PRODUCT INCLUSION TEAM

ANNIE JEAN-BAPTISTE

WILEY

Published by John Wiley & Sons, Inc., Hoboken, New Jersey.
Published simultaneously in Canada.

For general information on our other products and services or for technical support, please contact our Customer Care Department within the United States at (800) 762-2974, outside the United States at (317) 572-3993 or fax (317) 572-4002.

Wiley publishes in a variety of print and electronic formats and by print-on-demand. Some material included with standard print versions of this book may not be included in e-books or in print-on-demand. If this book refers to media such as a CD or DVD that is not included in the version you purchased, you may download this material at http://booksupport.wiley.com. For more information about Wiley products, visit www.wiley.com.

Library of Congress Cataloging-in-Publication Data:

Names: Jean-Baptiste, Annie, author. | John Wiley & Sons, publisher.
Title: Building for everyone : expand your market with design practices
   from Google's product inclusion team / Annie Jean-Baptiste.
Description: Hoboken, New Jersey : John Wiley & Sons, Inc., 2020. |
   Includes index.
Identifiers: LCCN 2020021970 (print) | LCCN 2020021971 (ebook) | ISBN
   9781119646228 (hardback) | ISBN 9781119646242 (adobe pdf) | ISBN
   9781119646235 (epub)
Subjects: LCSH: Product design. | Universal design. | Design–Social
   practices.
Classification: LCC TS171 .J43 2020 (print) | LCC TS171 (ebook) | DDC
   658.5/752–dc23
LC record available at https://lccn.loc.gov/2020021970
LC ebook record available at https://lccn.loc.gov/2020021971

COVER DESIGN: PAUL MCCARTHY
AUTHOR PHOTO: © SHAMAYIM SHACARO

Printed in the United States of America

SKY10020051_072420

To God, thank you for a multitude of blessings, including the opportunity to share these learnings and to continue learning throughout my life.

To all the beautifully underrepresented people of this world who live their truth in whatever manner they so choose, I see you.

To mum, daddy, and Herc, thanks for always being the strongest support system and for showing me that family will always have your back no matter what. Thank you for sacrificing, thank you for pushing, thank you for showing us that the American dream is possible but more importantly that the Haitian essence is undeniable. I will work every day of my life to make you proud and make sure that your sacrifices are worth it.

To Allan, my absolute best friend—thanks for believing in me and pushing me, always.

To my grandparents and ancestors, both here and in the stars, that have shown that Haitian and African spirit is resilient, bold, beautiful, and brave.

To Todd, my angel, my biggest champion. Thanks for always making me feel seen no matter what.

Thank you. I am eternally grateful.

CONTENTS

FOREWORD

The purpose of any organization is to build great products (or services) that solve people's problems or enhance their lives in some way. That mission is accomplished best when the focus is on the customer, whose needs, desires, and potential inspire innovation. Unfortunately, in the world of product design, we often get so wrapped up in design and product that we forget about the people who buy, consume, wear, drive, and use our products in other ways. This oversight is especially problematic, and common, when we are building for people who are different from us in terms of race, ethnicity, age, gender, abilities, language, location, and more.

In *Building for Everyone*, Google's Head of Product Inclusion, Annie Jean-Baptiste, brings the focus back to customers while widening the lens through which we view them. The resulting perspective, which is more inclusive of the vast diversity of the human population, enables and drives us to build products to meet the demands of a much broader consumer base. By taking a more inclusive approach to product design, development, and marketing, organizations stand to reap the benefits of increased innovation, customer loyalty, and growth. Along the way, organizations discover how to do well by doing good.

Understanding the phrase "product inclusion" needs to start from understanding Silicon Valley's use of the word "product." How that word gets used in Silicon Valley will often differ from the rest of the world, but it's becoming increasingly more salient in today's world. These days "product" is becoming more common as more people start to make products and services across industries. Getting comfortable with what Valley folks refer to as "product" will help you get into the right mood for reading this book, though the best practices span across industries.

A digital product is an app on your mobile device, or a website you click through, or a conversational bot you talk with. Techies who make these digital experiences refer to them as "products" in the same way that a baker will refer to their product as bread or a furniture maker will refer to their product as a chair. When you can see, smell, or feel bread with your senses and you can touch, kick, or trial a chair with your body, as products go they are real and relatable.

Digital products can feel like a far distance from a loaf of bread or a chair, but from a technologist's perspective they're indistinguishable. It's an unnatural leap to make because there's no product to point at when interacting with a mobile app or a website or amidst a conversation with a voice assistant. But when you make the switch happen in your mind, something amazing happens: you shift to a business view of the world where completely digital products occupy a bizarre world where marginal costs, inventory costs, and distribution costs go nearly to zero for a product. That's why both digital and physical products need to prioritize this type of work.

Now that you've started to take the ambitious leap to grasping the implications of wholly digital products, you can then move into the similarly disorienting depths of the word "inclusion." This shift requires you to decidedly jump in the diametrically opposite direction over to the non-techie, humanist perspective. Inclusive design expert Kat Holmes provides one of the best definitions of inclusion as simply: the opposite of exclusion. Have you ever experienced being excluded from something? Like a birthday party? Or a promotion? How did it feel? Hint: BAD.

Back to the phrase "product inclusion" and putting the two words together, we can read in this two-word phrase the juxtaposition of a new industry being defined by Silicon Valley that seeks to do the opposite of excluding others. That's because there's a history of tech products that have been unconsciously excluding others that you may not have known about. But more importantly, you're also going to learn how inclusion is now being introduced and prioritized into the digital bloodstream of not only tech products, but across industries like medicine, fashion, and more.

There was a time when the tech industry rallied around the mantra of "move fast and break things"—in other words, there was no need for accountability when the users numbered a few thousand people. But today we've interconnected millions and now billions of people. So to flippantly break things is no longer an acceptable outcome. The next generation of product creators, regardless of industry, subscribes to the newer belief of "move fast and ★fix★ things."

This book is filled with countless recipes to fix many of the experiences that we've now deployed at scale. Product inclusion will be the central challenge for any industry that creates a product or service to rally around—and this comprehensive guide provides the first atlas for navigating its many previously uncharted spaces. I know that in the many digital experiences I am charged with guiding today across industry types, there's going to be a pearl of wisdom from this collection that lets me put product inclusion first.

If you are in an industry other than tech, you may wonder what you can learn about product inclusion from somebody at Google who is so far distant from what you do. After all, a product in Silicon Valley is an app, a website, a search engine, a conversational bot—digital products that occupy a bizarre world in which marginal costs, inventory costs, and distribution costs drop to near zero. Just as Annie encourages you to widen your lens, she has widened hers for this book by presenting insights from a wide variety of business leaders in medicine, fashion, entertainment, fitness, and more.

I encourage you to not only read this book, but put it into practice. This book will serve as your guide to harnessing the power of diverse perspectives to drive innovation, growth, revenue, and more.

John Maeda
EVP/Chief Experience Officer, Publicis Sapient

INTRODUCTION

Being human is looking so deep within you that I see myself.
—Inspired by Mark Nepo, Poet and Spiritual Adviser

Think about a time when you felt completely yourself. Maybe you were with family members or friends that completely accepted you, or you were performing a hobby that you enjoyed immensely and could lose yourself in. You felt at ease, almost blissful, void of judgment because you could shake any preconceived notions of who you are, who you are supposed to be . . . and just be.

Now, think about a time when you felt excluded or ignored. Have you ever shown up at a party, summer camp, or a new job only to discover that everyone had already made friends and wasn't particularly welcoming or eager to get to know you? How did that make you feel? At the end of the day, everyone wants to feel welcome. It's what people want in their relationships with their family members, their friends, their pets, and their colleagues.

When we don't fit in or when products or services do not feel as though they were built for us, we feel excluded, frustrated, disappointed, or even upset. When a product or service seems to be designed for everyone but us, we can feel as though we were ignored or disregarded by those involved in the product design process. Our feelings can range from mere annoyance ("Whatever, I didn't want to use this thing anyway") to deep alienation or hurt ("I feel like this does not represent me or my community and what this represents is threatening").

Knowing that we've all had an experience of being *othered* (marginalized by a social group that considers itself superior) regardless of our background, it's imperative that we don't create that feeling in people who interact with our products, services, content, marketing, or customer service. When creating products, we want to avoid building anything, even unintentionally, that makes anyone feel this way. One of the goals of this book is to help you and others in your organization avoid engendering this feeling in the users of your products or services.

As designers, creators, engineers, user researchers, marketers, and innovators, we want everyone to feel included. Isn't that why we got into this work—to be able to create products, services, and content that shape the world for the better, that empower people to live richer lives, to experience things they haven't with the people (or creatures) they love? At the heart of this commitment is inclusion: everyone seeing themselves in the end result of a company's or an individual's work. People want to feel seen, heard, and considered; they want to feel that people like them matter to companies, that their unique backgrounds and perspectives are valued.

It's not enough to *want* to be inclusive. We must think and act with intent and deliberation. We must center inclusion at key points in the design, development, testing, and marketing processes to ensure that differences among users are considered and addressed. As diversity and inclusion champion Joe Gerstandt[1] reminds us, "If you do not intentionally, deliberately and proactively include, you will unintentionally exclude." This phrase is often repeated at Google to remind our teams that merely wanting to do the right thing isn't enough.

Planning for Success

The adage "Failing to plan is planning to fail" definitely applies when you're designing for inclusion. Because thinking about diversity, equity, and inclusion in product design may be new to you or your team, having a solid plan that everyone understands and buys into is important.

As with any other endeavor, well-defined roles, deadlines, objectives, and metrics are key components to successful implementation or execution. We touch on each of these topics in this book as we put theory into practice and provide guidance on how to bring inclusion into the various phases of product design, including ideation, user experience and design (UX), user testing, and marketing.

---

[1] www.joegerstandt.com.

As we plan for inclusion, following the golden rule—treating others as we would like to be treated—isn't the way to go. We need to adopt the platinum rule—treating others as *they* would like to be treated. Brian Stevenson, Executive Director at the Equal Justice Initiative (EJI) and author of *Just Mercy*, encourages us to get "proximate." *Getting proximate* is the act of getting closer to someone, to understand their experiences, their fears, their hopes. The goal is to build empathy, which hopefully leads to action.

Although Stevenson discusses the need for empathy in a different context, this concept is very applicable to product inclusion as well. As a whole, businesses and other organizations haven't been proximate with all consumers, including people of color, lower socioeconomic status, and advanced age; people who live in rural communities or outside the business's home country; people with disabilities; and those who are members of the LGBTQ+ community. When dimensions of diversity intersect (for example, a Black woman over 50), the challenge to serve consumers' needs and preferences becomes even more nuanced. Becoming proximate enables you to understand, to build empathy, to want to be better, and to want to *do* better. It provides the drive to hold yourself, your teammates, and your organization's leadership accountable to truly building for users regardless of their background.

However, wanting and doing are light years apart, and that is where planning enters the picture. Planning is the bridge that connects the two. Planning enables organizations to break deeply ingrained thinking and behaviors. Planning holds the hope and promise of changing the culture.

You may be thinking, "OK. I get it. I can't only *want* to build inclusively. I have to *plan* to build inclusively. And then I need to *do* it. So tell me how!" We will get there. But I want to be sure that I ground this book in how integral inclusion is to building and marketing products that truly resonate with your customers and users, because this work can get messy, complicated, frustrating, and awkward. For example, when you're ready to launch, you may discover that your product's colors are not discernable to someone who is color-blind. Or, as your marketing team is putting the final touches on its latest campaign, someone points out that all the people of color depicted in the ads have lighter skin. Or, you are getting ready to launch in Latin America only to find out that your translations from English failed to account for the cultural nuances of different Latin American countries. Setbacks such as these can derail the best of intentions; having a plan keeps everyone on track and the earlier on in the process you bring product inclusion in, the more likely you can avoid these challenges and find even more opportunity.

Making Inclusive Design a Priority

I understand that all organizations have priorities and product inclusion, resource constraints, and time constraints. I get it—you're probably strapped for time and cannot imagine adding yet another initiative to your workflow. Perhaps you have heard about product inclusion already and bought into it as the right thing and the best thing to do, and you may think you already have a pretty good idea of who your user is.

I understand because I've been there—I've worked with hundreds of teams and businesses, from small and medium to large organizations inside and outside of Google, helping them with their holistic advertising strategies on Google's platforms. I've sat with them and heard their concerns and helped them grow their businesses and think up new ones. I've consulted businesses on resources and understand that organizations must be ruthless in prioritizing in order to succeed.

As I was writing this book, I took into consideration that organizations have other priorities and often limited resources. I accommodated those challenges by breaking down product inclusion into four main phases—ideation, user experience (UX), user testing, and marketing—and providing the option to start slowly with only one or two. Early in this book, I offer advice on how to get started even if you have a small team by engaging with more diverse users; for example, you may want to talk to real users and leverage their stories in your marketing. Later in the book, I introduce tactics and techniques that cross all four phases.

However you decide to start, realize (as I and my team have realized) that product inclusion is an exciting, fun, and never-ending journey of discovery. My team hasn't always gotten it right. We are learning together and hope to learn with others who are doing great work in this space. We also are learning from others who have committed to product inclusion practices in the hope that, together, we can create an ecosystem and share best practices across industries to create more inclusive products.

We are excited about the journey ahead and we know that in order to serve billions of users across many dimensions, including race, ethnicity, ability, sexual orientation, gender, socioeconomic status, age (and more!), and across the intersections of these demographics, we need to prioritize and maintain focus on inclusion.

Across Google, when people talk about diversity, equity, and inclusion, we often liken it to going to the gym and building a muscle. At first, you may dread the challenge and the effort required, but as you build your product inclusion muscle, it gets easier, and the easier it gets, the more fun and exciting it becomes. You'll be able to do more, you'll feel yourself growing more confident, and you'll look back on your journey and be proud of what you and diverse others have accomplished working together. Don't think that just because you may not get it all the first time (or the first couple times) that you have failed; every failure is a learning experience often accompanied by new, unforeseen opportunity. Thinking and talking about inclusion are fantastic first steps. After all, ideas and conversations are the seeds of innovation, and when those seeds come to fruition, your users will thank you, and your organization will prosper!

Yes, your organization *will* prosper. Many people mistakenly assume that underrepresented users comprise an insignificant portion of the population, so making them a priority is a low priority business decision. That assumption and the conclusion on which it is based are incorrect. If you subscribe to this mistaken belief, I urge you to shift your thinking to think less about who your users *are* and more about who they *could be*. Draw your circle a little bigger to encompass those standing outside it. As you do, you will begin to notice people who may not look, act, or think like you, but like you, they are yearning to feel seen through the products and services you offer. They may represent another gender, race, or socioeconomic status or a combination of these dimensions. Their voices may not be those traditionally heard and listened to in the product design process, but theirs are the voices that will define the future or your products, making them richer and better overall.

As you expand your circle, you bring consumers who are unserved and underserved into your compass. You uplift their needs and make them core to your practices and your processes. They become both passive and active participants in your design process. As you prioritize

Proactive inclusion in product development can help us design useful, accessible products for large, untapped audiences.

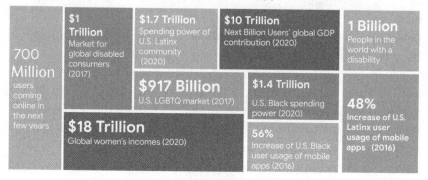

Figure I.1 **Opportunity and purchasing power for underrepresented demographics.**

inclusive design, you begin to center on deeply held user concerns with the aim to resolve those concerns, which is fantastic for business and essential for companies to remain relevant now and in the future.

If investing in product inclusion seems risky, consider the risk of ignoring the potential just in terms of revenue and profits. A ton of opportunity is on the table, as shown in Figure I.1.

This is groundbreaking. By building with the diversity of the world in mind and centering inclusion in your product design and marketing processes, you have the opportunity to tap into the trillions of dollars of global spend being left on the table by less enlightened organizations.

If you need additional proof or you believe that "the proof is in the pudding," take a look at how we are integrating product inclusion at Google—the subject of the following section.

Bringing an Inclusive Lens to Google Assistant

At Google, our inclusion team plays an integral role in the product design and development process. As partners in the process, we want to ensure that all Google Assistant–enabled products deliver an inclusive experience, and our partners, Beth Tsai and Bobby Weber, share our commitment. They were proactive and intentional about bringing inclusion into the process before launch, as opposed to identifying and resolving issues later in the process, which is far more difficult and costly. They wanted to delight customers and felt that applying an inclusive lens to the process was a key factor in doing so. The team wanted to ensure that at launch, Assistant did not offend or alienate users based on race, ethnicity, gender, sexual orientation, or other dimensions that make our users who they are.

We realized that in order to make Assistant inclusive, we had to bring in a multitude of diverse perspectives. We worked with the team to "stress test" (adversarially test) Assistant. We brought Googlers from different backgrounds and perspectives together in war rooms to try and break the assistant using the cultural background they brought to the table. We knew that Googlers from one of our *affinity groups* (a group of individuals linked by common interest, purpose, or diversity dimension) would have more expertise than Google Assistant or a small team of developers would have in respect to what certain communities would find alienating or offensive. We also knew that communities are not a monolith, and so asking one person would not be representative of a whole community.

Our product inclusion champions tested Assistant based on the assumption that some users were likely to issue racist, sexist, homophobic, and otherwise offensive questions and commands. For example, as a result of our efforts, if you ask Google Assistant "Do Black lives matter?" it says, "Of course, Black lives matter."

By integrating inclusion in product design and development, we significantly reduced the number of escalations requiring action at launch. (An *escalation* is an act of exploiting a bug or design flaw in a product.) Escalations can hurt your brand, erode user trust, and slow sales. All of this is detrimental to a business.

At launch, Assistant had .0004 percent of total interactions that needed to be acted upon. In other words, out of billions of queries at launch, only .0004 percent of queries were so egregious that they needed to be acted upon. This was a huge success and very important given Assistant's growth and reach. In addition:

- "Available in more than 90 countries and in over 30 languages, Google Assistant now helps more than 500 million people every month to get things done across smart speakers and Smart Displays, phones, TVs, cars and more."[2]

- "Google Assistant is already available on more than 1 billion devices."[3]
- "Active users of Google Assistant grew four times over the past year."[4]

Even with Assistant's growing popularity, we have seen minimal escalations, partially due to the fact that we prioritized product inclusion and integrated it in the design, development, and testing processes.

Of course, surveying every single person from every single community before launching a product is impossible, but bringing diverse perspectives to the table is imperative. You are probably already conducting focus groups and performing other types of user research; making these processes more inclusive is simply a matter of increasing the diversity of the researchers and participants.

Building for Everyone, with Everyone

At its core, this book is about people with a diversity of backgrounds working together to build and market products and services for everyone regardless of race, color, belief system, sexual orientation, gender identity, age, ability, or other qualities that make us different. Our credo (coined by my former teammate Errol King) is to "build for everyone, with everyone." That commitment is what our team and what many teams at Google are working to instill intentionally and thoughtfully into the people and processes that create our products and services. Our goal is to ensure that no matter who you are or where you live, you can find benefit and delight in our products and services.

This book doesn't sugarcoat the complexity of addressing the needs and preferences of our incredibly diverse human population, and it doesn't underestimate the time and effort required to become proximate with the multidimensional and multifaceted nature of individuals. What this book does is help you appreciate the value and necessity of placing

---

[2]https://blog.google/products/assistant/ces-2020-google-assistant/ (January 7, 2020).
[3]Ibid.
[4]https://bgr.com/2019/01/07/google-assistant-1-billion-devices-android-phones/ (January 7, 2019).

human beings at the core of your design principles and practice, and it shows you how to do just that.

We are living in a very exciting time, a time when the historically underrepresented and disenfranchised are empowered and empowering themselves to wholly participate and lead in all aspects of life, including the global economy. These groups, like all people, deserve to be seen, heard, and served by the products and services we offer. We need to build for everyone, with everyone, not only because it is the right thing to do, but also because it drives innovation and growth while making the world a better, richer place.

Product inclusion basically boils down to listening, caring, and being humble. We're all on a journey toward happiness and self-fulfillment, traveling along a path strewn with challenges and opportunities. We can help one another along this path and tap into the opportunities by doing so. Just put yourself in the shoes of consumers (be intentional about bringing underrepresented voices to the fold), ask questions, consider the problems they face, the opportunities that stand before them, and how they may be currently excluded from the solutions and benefits available to others, and keep trying. The opportunity to build better, more inclusive products and grow your business in the process awaits!

This book will cover learnings that I've been fortunate enough to help shape, and I'm continually inspired by the community that continues to make product inclusion even more of a priority, both inside and outside of Google.

When we succeed in building and delivering inclusive products and services, we will begin to fully reflect the beauty and diversity of our world and to prosper in all ways. We will continue to identify and create new markets and build wealth, and as we lift the entire population through our efforts, we, too, will reap the benefits and experience the fulfillment of having made a positive impact in the lives of others.

CHAPTER 1

Building for Everyone: Why Product Inclusion Matters

Across industries, we've heard that inclusion is "the right thing to do." Good people and ethical organizations believe in and practice diversity, equity, and inclusion. This is certainly true, but it oversimplifies the reason why people and organizations should care about and embrace these virtues.

Diversity, equity, and inclusion matter for two reasons. The first is a human reason—people matter. Diversity enriches the world with different languages, perspectives, customs, foods, clothing, art, innovation, and much more. Equity and inclusion are essential for making people feel welcome, appreciated, and empowered; in a way, they enable everyone to flourish and to contribute in all ways that make them unique.

The second is a business reason—diversity, equity, and inclusion are good for business and for all productive human endeavors. Organizations that engage with people representing a wide range of demographics reap ideas and innovations that vastly improve their products and services and even open their eyes to new markets and entirely new businesses. As a result, they grow their customer base, increase innovation, and build momentum over less inclusive players.

To create for the world we live in, we must build in an environment that reflects that world. We cannot build for people without understanding them, their needs, their preferences, and what disappoints and upsets them and makes them feel excluded. The world is changing, and this change is accelerating. You can witness it happening around you—in the news and entertainment programming you tune in to, the advertisements you see, and hopefully in your neighborhood and workplace.

I encourage you to embrace this change, and I challenge you to take the lead in promoting this positive transformation by creating products and services that represent the makeup of this ever-evolving world. The first step is to develop an understanding of your users (customers or clients)—who they are, where they come from, what's important to them, and how their core needs align with your organization's and your work's mission. Having this understanding is the key to unlocking value, thereby opening the doors to growth and innovation.

Absence of inclusion is a threat to business

—Daisy Auger-Dominguez, Founder, Auger-Dominguez Ventures, a workplace culture consultancy and advisory firm DaisyAuger-Dominguez .com

The notion of building inclusively is one that the tech industry has lifted up, but that any company—from one producing consumer products to one creating content—must take seriously. Why? Because of an increasingly diverse customer and client base who embrace and value authenticity in their products, services, and content diversity. Add to that growing shareholder, employee, and customer activism, and leaders and companies are at risk of surviving if they fail to build a diverse workplace and adopt inclusive business practices.

For the most part, companies tend to be reactive in this area, taking action only in response to loss of market share and diminishing brand loyalty and financial returns. But with increasing competition across every industry, companies can no longer risk waiting to create products that more closely resemble, serve, and connect with their customers, clients, and viewers.

From the lens of service-based organizations, we can expect an increasing number of prospective clients to ask the question, "If you can't run your business inclusively with a lens on the landscape of today and the future of work, why am I trusting you to advise me on what to do?" That's a fair question and one I believe will surface time and again.

If voluntary change does not happen soon enough, it will happen by uncomfortable client-led, employee-led, and investor-led forces. Bottom line: Companies that fail to build safe, inclusive, and diverse workplaces and adopt inclusive business practices risk their bottom line, brand, and reputation.

Product Inclusion Team Approach

On our diversity and inclusion site, we say "Google's mission is to organize the world's information and make it universally accessible and useful. When we say we want to build for everyone, we mean everyone. To do that well, we need a workforce that's more representative of the users we serve."

When we build products at Google, we think in terms of building for everyone, with everyone. This is hard work. We all have biases and can't understand every culture, preference, and individual need. We've all made mistakes and will likely continue to do so, but when we commit to striving to do better, to bringing in different perspectives, to calling out bias, and to owning our mistakes, we can begin to design and build more inclusive products. It's a journey, and we're committed to learning and improving.

While our work focuses squarely on the human factor and is representative of work being done not only inside Google but within many other companies, tech and non-tech, it is decidedly unique in that it is deeply rooted in product and marketing. It also is grounded in business metrics and data around improving the bottom line across multiple dimensions of diversity and the intersections of those dimensions. (We'll dive into the capstone research our team did to prove the business case for inclusion in Chapter 2.)

At Google, we also work hard to get everyone, at every level of the organization, involved in the process, and I recommend that you do the same. Whether you are a business leader, product manager, program manager, marketer, or designer, or you work in an industry that isn't tech, you can contribute significantly to the success of your product or service by leveraging insight from your own background and experience and by viewing your work through a lens of inclusivity.

Thinking inclusively: An engineer's perspective

—Peter Sherman, Google Engineer

Inclusive thinking can begin with an exercise that should be simple for most engineers—consider that you (the engineer) are often the technical support specialist for your non-technical friends and family. As a technically minded individual, you view many of their questions/issues as simple, and you solve them easily. In that moment, it can be helpful to understand what the experience is like on the other side—to use a product or service that is seemingly not built for you. Consider that for any consumer product, "built by engineers for engineers" is probably not a good product development strategy—at best, it will limit the audience; at worst, it will alienate users and make them feel unvalued. Non-engineers may not understand the design and the process flow, or may be confused by the instructions and may abandon the product altogether.

By considering the entire potential audience for a product, engineers can design features to work well for all of their users. We commonly apply these principles with general user experience (UX) and user interface design and with accessibility and internationalization/localization. But to truly design for everybody, we must think even more broadly. Ask if there are inherent biases in a product, service, or system. For example, a camera might have bias in the color rendering of skin tones; an app may be built with connectivity requirements that can be satisfied only in certain communities; a form may require inputs in a specific character set; a service may assume a certain level of digital literacy and access that is exclusive to certain users; the list goes on.

Whenever I approach a new system, I try to consider how a user will interact with a feature or product, what inputs are relevant, how the system works with the data and how it perceives the world, and ultimately what underlying technologies, assumptions, and tuning/learning will contribute to the final output. With this in mind, discussions with other teammates, managers, and users—especially those who are not like me and can bring a different perspective to the table—can uncover potential things we've missed with respect to possible built-in bias and how we can address those issues to make the products as inclusive as they can be.

Above all, I believe that the engine for creating truly inclusive products is the collaborative engagement with people who have diverse backgrounds and perspectives—when we work together, we form the lever that can move the world.

Understanding Who Your User Is

All organizations attempt to understand the users of their products and services. They need that understanding to evaluate marketability, but they also employ their user analysis in the design and development of products and services. This practice is nothing new. What is new is the increasing necessity to consider differences among individuals in order to serve a broader consumer base.

When done right, inclusivity is baked into the product. In tech, for example, people of color are underrepresented. To build inclusivity into a new tech product, we need to bring people of color into key points in the design, development, testing, and marketing processes to reduce bias. Consider another example. Suppose a team wants to build a camera that can be used for home monitoring and security. Many teams begin by defining their user along with their core challenge. They decide who is most likely to use this product and identify what core need this particular user has. Suppose the team decides they are building this app for moms who want to be able to watch over their homes when they are at work or school, running errands, or away from home for other reasons. The team created a very specific target user, which isn't necessarily bad, but it does limit the market and potentially leaves a lot of money on the table. The team would be wise to broaden its scope to include stay-at-home dads/parents, grandparents, caregivers, and business owners, and the diversity these different users represent could significantly impact design decisions related to features, functionality, and more.

Unfortunately, companies often build products for a small subset of people familiar to and often similar to the people creating the product; this is called the "like me" bias. What happens is that a majority group frames the persona of the target user and the core business challenge; as a result, they fail to include people who have been historically underserved by the industry overall, whatever that industry happens to be—tech, finance, fashion, entertainment, and the list goes on. An example of this might be picking a name for a product and not realizing that it means something completely different, or even offensive in another language.

The further along in the process you are before you bring in these users and understand their needs, the harder and more expensive it is to build an inclusive product. Even if you succeed to some degree, products that are built for inclusion as an afterthought often carry an air of inauthenticity. Historically underrepresented users can often sense when they were not thought of early in and throughout the design process.

Product teams often begin with a specific user in mind. What product inclusion helps to do is expand the scope of the target user to unserved or underserved populations that may benefit from the product or service. Starting with a narrow demographic is okay, but product teams, with deliberate intent, must learn to broaden their scope and think through who else could use the product and in what circumstances or situations.

Everyone has bias. Everyone is susceptible to looking past people in certain demographics, so teams need to be intentional about bringing different perspectives to the table. Ideally, you want input from people representing diverse dimensions and the intersection of those dimensions, but if that is not an option, then at least pull in colleagues from another team or department; for example, if you are in product development, invite someone from marketing or human resources. Brainstorm together, challenge assumptions, poke holes in one another's reasoning. Let the process get messy. Remember, the purpose is to reveal bias and uncover potential areas of exclusion.

When you expand to different demographics, you begin to notice other functionality or applications you hadn't considered. New opportunities become apparent because different people have different perspectives, different use cases and needs, and different preferences. Bringing together diverse perspectives results in richer, more innovative end products with broader appeal.

When we discuss inclusion in tech circles, people often say, "If you can see it, you can be it," when talking about diversity and inclusion, meaning if you can imagine someone similar to you in a specific job, you will consider that job attainable for yourself. Products and services often reflect their creators, and you can usually tell, consciously or sub-consciously, by looking at the product or service or using it, whether the creators were like you. When people are reflected in your mission, vision, product, and marketing, they sense that they were the target user or that this was something that was built with them in mind. Your goal in product inclusion is to enable as diverse a consumer base as possible to see their reflection in your company, your products, and your marketing.

Understanding What Exclusion Looks Like

Technology can and should be a great equalizer. All around the world, the Internet connects people with information that can change their lives for the better, reveal what is possible, satisfy their curiosity, impart knowledge and skills, and bring them closer to self-fulfillment. It provides information and resources that can be accessed anywhere at any time on demand. But millions of people don't have access to the Internet, so they can't benefit from its rich resources and, perhaps more importantly, can't contribute to the shared knowledge found online.

Suffering from a lack of Internet access is only one example of exclusion, but considering this example can help to develop empathy for people who are overlooked or ignored. It provides a glimpse of what exclusion looks like and, perhaps more importantly, what it feels like. Shining the light on exclusion is important, regardless of how uncomfortable it may make us feel, because exclusion is the opposite of inclusion. Just as we can only fully appreciate light by knowing darkness, we can only fully understand inclusion by understanding exclusion and empathizing with how it makes people feel.

What does it look like to exclude a user who cannot afford your product? Or one who cannot understand the English-only instructions shipped with your product? What about someone with a vision impairment who needs a screen reader to use your product? Understanding what it feels like to exclude and feel excluded is important to ultimately adopting an inclusive mind-set and doing the hard work of making products more inclusive.

User exclusion usually is unintentional. Rarely, if ever, does a product developer make a conscious decision to design a product unsuitable for a certain demographic. Exclusion can happen often as a result of who gets assigned to the product teams. Because we all have bias, we deceive only ourselves when we think we understand every user's needs and preferences. But just because our decisions or actions are unintentional does not lessen their impact on people who feel excluded.

Historically in tech and in many other fields, product team composition has not been representative of the world at large; teams did not look like the users of the products and services they were building. To become more inclusive, we need to build diverse teams and ask users what they need. We need to sit down with users, see how they interact with the product, observe how they live, converse with them openly throughout the process to identify their core challenges and gather their input and ideas, and then adapt and build. This process is difficult without a diverse group of people making key decisions about what to build, how to build it, why to build it, and for whom, which provides even greater impetus to put infrastructure and processes in place to encourage and facilitate product inclusion.

The benefits are well worth the effort—increased innovation, customer satisfaction, growth, and more. Doing good for customers and doing well as a business are not mutually exclusive. In fact, the two have a synergistic relationship. It is becoming increasingly clear that diversity and inclusion are crucial for running a thriving business. In addition, product inclusion does not necessarily add to anyone's workload. When diverse teams are in place and everyone in the organization has adopted an inclusive mind-set, inclusion is integrated seamlessly into processes and can be a great amplifier of the work.

Putting technology in the hands of underrepresented users

—Daniel Harbuck, Global Social Impact and Sustainability Program Manager

As Global Social Impact and Sustainability Program Manager, in 2019 I worked with a dedicated team to deliver Google Home devices to people in need. Specifically, Google Nest partnered with the Christopher and Dana Reeve Foundation to bring a Google Nest mini into the homes of up to 100,000 people living with paralysis.

Our team is focused on disproving the assumption that underrepresented groups are "edge cases" and to present indisputable evidence that these groups are actually the solutions that improve life for everyone.

Any little thing we could do to empower these voices, from design to delivery, felt right.

We already knew that technology has tremendous potential to help bring people closer together. From self-driving cars to mobile phones and wearables, like earbuds, innovation brings a welcome improvement to the world of traffic safety, family connectedness, and personal health.

In practice, unfortunately, we let technology separate us. One in five people live with a disability, which, by almost every measure, constitutes the largest underrepresented community in the world. It is as if voice-activated smart home devices were built for people living with paralysis, but, in practice, people living with disabilities are 20 percent less likely to have technology in the home (http://pewrsr.ch/2og9Q4z).

This group of people living with paralysis is not an edge case to support in the next version; these are precisely the people our products need to be engineered to help from day one. Every

(continued)

(*continued*)

time we design for greater independence for a person living with disability, we also improve the lives of others often in unexpected ways. Improved voice recognition technology, driven by privacy-protecting machine learning, makes it easier not just for people with disabilities to more accurately and swiftly connect with loved ones, simplify their daily routines, and enjoy favorite entertainment: it improves the product for everyone.

Speaking a Common Language

Diversity in language can lead to misunderstandings, so defining key terminology is important. At the core of product inclusion discussions are four key terms—*product inclusion, diversity, equity,* and *intersectionality.*

Product inclusion is the practice of applying an inclusive lens throughout the entire product design and development process to create better products and thus, accelerate business growth. It involves bringing in diverse perspectives at key inflection points in the process, then embedding them in the process to create products for a more diverse consumer base.

Diversity is variation within a group of people; this variation includes differences in social identities (gender, race, ethnicity, age, sexual orientation, gender identity, ability, class, socioeconomic status, etc.); background or personal attributes (education and training, experience, income, values, worldview, mind-set, faith-based affiliations, etc.); and other differences (location, language, available infrastructure, etc.)

Equity is the quality of being fair and impartial in terms of access, opportunities, and success for all individuals. Equity requires that we remove any predictability of success or failure that currently correlates with any social or cultural identity markers. Do not confuse equity with equality. *Equality* implies the same for all, whereas *equity* provides what each individual needs to be successful. A focus on equity intentionally interrupts unfair practices and eliminates bias.

Intersectionality is a term coined by Professor Kimberlé Crenshaw and is defined as "the interconnected nature of social categorizations such as race, class, and gender as they apply to a given individual or group, regarded as creating overlapping and interdependent systems of discrimination or disadvantage."

It's also important to note that individuals are often diverse across multiple dimensions. For example, I am a Black, nearly 6-foot-tall, first generation Haitian-American, left-handed woman. I speak multiple languages, am an introvert, sometimes mix up numbers and letters or read things in the wrong order, and am a very visual learner who prefers learning by doing over being taught. Those dimensions cover race, physical dimensions, learning styles, and more, all of which influence how I move in the world and how people perceive me, yet they are a small subset of what makes me different from others. My qualities must be considered as a whole; I am not Black on Monday, first generation Haitian-American on Tuesday, and left-handed on Wednesday. All my dimensions are constantly in play, and they influence how I use and am connected to products and services.

Having diverse perspectives is not enough; we must allow our differences to be seen and to develop. When we talk about and accept inclusion, we demonstrate that we value the perspectives and contributions of all people and intentionally consider and accommodate their diverse needs and viewpoints. Inclusion brings about the following:

- A culture co-created by every member of the group, not merely an invitation for people from historically underrepresented groups to be present in an existing space.
- A culture that enables different people to belong and operate in authentic ways.
- A culture that ensures all people are treated respectfully.

With product inclusion, we are adding products and services to the conversation of diversity—thinking and talking about diversity, equity, and intersectionality in the context of product design and development. I would go further to argue that these terms, concepts, and practices are an integral part of product design and development because no product is perfect without integrating inclusion into the process.

Expanding Diversity and Inclusion to Products and Services

Traditionally, diversity and inclusion have focused on culture and representation within an organization. Product inclusion expands on that notion and focuses not only on building inclusively, but also on framing diversity and inclusion not only as "the right thing to do" but also as a sound business practice. As an old adage advises, you can "do well by doing good," and one key way that organizations can do good is by treating people equitably in their hiring practices and in their product design and development processes, which go hand in hand.

The concept of doing well by doing good is evident in the data. In August 2019, the Female Quotient partnered with Ipsos and Google to survey 2,987 U.S. consumers of various backgrounds to better understand perceptions surrounding ads they consider to be diverse or inclusive.[1] According to Virginia Lennon, Ipsos Senior VP of the Multicultural Center for Excellence and one of the lead researchers on the study, "The purpose of this study was to help us better understand how consumers see authentic representation in ads, images, and within organizations."

The study asked participants about their perceptions with respect to 12 categories related to diversity and inclusion in advertising. Specifically, they were asked to think about which of the following, if any, they believe are important for brands to be conscious of in order to be inclusive and diverse in their ad campaigns: gender identity, age, body type, race/ethnicity, culture, sexual orientation, skin tone, language, religious/spiritual affiliation, physical ability, socioeconomic status, and overall appearance.

Then, the participants were asked about what actions, if any, they have taken related to a product or service advertised after seeing what they considered to be a diverse or inclusive ad campaign. The eight "product-related" actions people could select were:

- Bought or planned to buy the product or service
- Considered the product or service

---

[1] https://www.thinkwithgoogle.com/consumer-insights/inclusive-marketing-consumer-data/.

- Looked for more information about the product or service
- Compared pricing for the product or service
- Asked friends or family about the product or service
- Looked for ratings and reviews of the product or service
- Visited the brand's site or social media page
- Visited a site/app or store to check out the product

The study found that 64 percent of U.S. consumers said they took at least one of the eight different product-related actions after seeing an ad that they considered to be diverse or inclusive with respect to the 12 categories discussed in this study. And this percentage is higher among specific consumer groups including Latinx (85 percent), Black (79 percent), Asian/Pacific Islander (79 percent), LGBTQ+ (85 percent), millennial (77 percent), and teen (76 percent) consumers. Of the various groups surveyed, LGBTQ+ and Black consumers expressed the strongest preference for diverse and inclusive ads.

Overall, the results of the study reaffirmed that the general population and historically underrepresented consumers today are highly attuned to authentic representation in ads—and their rising expectations surrounding diversity and inclusion influence their choice of brands, products, and services.

Below are some additional insights from the research:

- 69 percent of Black consumers are more likely to purchase from a brand whose advertising positively reflects their race/ethnicity.
- 64 percent of Black consumers say they are more likely to buy from a brand that hires women, minorities, and underrepresented people to build its products or services.
- 71 percent of LGBTQ+ consumers are more likely to proactively seek out a brand whose advertising authentically represents a variety of sexual orientations.
- 68 percent of LGBTQ+ consumers would be more likely to purchase from a brand whose advertising positively reflects a variety of sexual orientations.
- 60 percent of LGBTQ+ and Black consumers say they think companies that hire women, minorities, and underrepresented people create better products and services compared with those that don't.

As Virginia says, "We now have generations of consumers who are increasingly multicultural through the intersectionality of race, gender, ethnicity, and sexual orientation. This study clearly told us that these consumers expect brands to be inclusive and reflect the reality of their lives in advertising."

Similarly, additional research we've conducted at Google shows that a majority of consumers overall want companies to prioritize inclusion (see Chapter 2).

Collectively, the data reveal a huge opportunity for businesses to meet consumers' rising expectations for diversity and inclusion. And because diversity and inclusion are interlaced with culture and representation and the fact that inclusive design produces superior products and services, the "right thing to do" is inextricably linked to the business case for diversity and inclusion.

Recognizing the Need to Communicate with Underrepresented Users

Product teams are driven by satisfying a user need or solving a user problem. Unfortunately, they often have a misconception of what customers need, the factors relevant to a problem, how the factors are related, and how a product might solve the problem or satisfy the need. These misconceptions often miss their mark by a degree directly proportional to the difference between the members of the product team and the customers they are trying to serve. Factors that contribute to these misconceptions include mistaken assumptions, narrow perspectives, and inherent biases of team members.

Additionally, product teams often fail to factor in the dynamic nature of the forces that define customers' needs and preferences. For example, several factors influence consumer behavior, including economic conditions, culture or group influence, purchasing power, personal preferences, and situational challenges. All these factors and how they interact are subject to change. The resulting dynamic, with hidden feedback loops that underlie even seemingly simple problems, can be very complex and difficult to understand, especially from the outside looking in.

As a result, hypotheses about what customers need or want are often merely best guesses that ignore the broader societal context in which products will ultimately operate. These hypotheses are typically neither transparent nor informed by the perspectives and lived experiences of peripheral stakeholders (customers). The resulting products often miss the core need or lead to unintended product outcomes, such as excluding an entire segment of the population, which is why involving a diversity of perspectives in the product design process is so crucial.

Awareness of the dynamic nature of problems and needs in complex systems (social, economic, ecological, managerial, and so on) is nothing new. In fact, over 50 years ago, it gave rise to the field of *system dynamics*—a methodology used to understand and discuss complex issues involving interdependence and mutual interaction of multiple factors, feedback loops, and circular causality. I learned about system dynamics in large part from two Googlers: Donald Martin and Jamaal Barnes, with whom I frequently collaborated across multiple projects. I spoke with Donald about what system dynamics means in the context of product inclusion (see the nearby sidebar).

Accounting for system dynamics

—Donald Martin, Jr., Technical Program Manager & Social Impact Technology Strategist, ML Fairness, Ethical AI and Trust & Safety

System dynamics (SD) is a methodology, invented at MIT over 50 years ago, for describing and modeling complex problems both qualitatively and quantitatively. It is optimized to contend with the feedback loops, non-linearity, and timedelays that characterize the dynamically complex societal context in which we build and deploy products. SD is a well-honed practice for making the implicit and incomplete hypotheses about user needs that drive product development, explicit via visual models called causal loop diagrams and stock and flow diagrams. Inherent to the approach is convening and facilitating group model building sessions in which hypotheses are transparently and collaboratively evaluated and refined from multiple diverse perspectives. A key artifact of the process is a shared dynamic

(continued)

(continued)

hypothesis of the problem to be solved that can be quantified and simulated to gain deeper understanding. Community-based system dynamics (CBSD) is a variant of SD, which is focused specifically on building capacity within communities to describe and model the problems they face directly, without intermediaries. The first-hand descriptions and models generated by these communities can be valuable sources of human-experience and user-need data for user experience researchers and product managers who are motivated to build products that work well for everyone, including those from traditionally marginalized groups. CBSD can enable typically excluded stakeholder communities to fully participate in the early stages of problem understanding and product conception as partners who bring valuable expertise versus afterthoughts.

The current machine learning (ML) revolution brings great promise but also great hazards. ML systems tend to amplify the biases that exist in society. As such, the stakes for understanding the societal context in which ML-based products will operate are even higher. CBSD is a promising approach for gaining the understanding required to build inclusive ML-based products with everyone.

To build for everyone, with everyone, teams should partner not only with users from different backgrounds but also with other stakeholders such as experts who are well versed in the problem domains targeted by the product, as well as the communities who have a stake in those problems. These additional partners and stakeholders can help teams better understand the implications of their proposed solutions and identify additional problem domains and target markets to explore.

For example, suppose a team wants to build a robot that picks up a customer's dry cleaning, drops it off at the dry cleaners, and then delivers it to the customer. The team reaches out to various friends, who emphatically let the team know that they would use the service. The team starts to build their prototype. By limiting communication to team members and "friendlies," the team is likely to have a limited perspective of customer needs and preferences. They would benefit from challenging their assumptions by asking questions such as the following:

- Are we missing any demographics in the people we consulted?
- What is the root challenge—convenience, cost, speed?
- Is it convenience; is it taking special care of certain clothes?
- Who is left out of a dry-cleaning app and is there a way to expand? For example, what about people who may only want their laundry done? What about the difference in prices of dry cleaning depending on your gender? What about bringing the lens of socioeconomic status?
- Is there actually a larger need for this product?
- How does this affect the environment and the communities (local dry cleaners)?

Before beginning, the product team should put their hypotheses in writing and test them not only with "friendlies" but also with people who may not approve of the idea. Feedback from those who may not see value in the product is critical, as this feedback may highlight weaknesses and poke holes in initial assumptions before the team gets too far into the design process, risking resources and ultimately product success.

The team should also proactively seek feedback from non-target stakeholders and communities that could be impacted by the product. A critical difference exists between seeking feedback to validate a proposed product intervention (for example, "We are building this dry cleaning app, what do you think about these features?") and seeking feedback on the underlying problem itself (for example, "We want to simplify and bring down the cost of cleaning clothes. What challenges or opportunities do you see, and what ideas do you have?").

When you sit down with people and really listen, you can unlock potential for positive social impact and market/user expansion. You may discover a previously invisible need/problem and find a solution that you would never otherwise have imagined, one that may appeal to hundreds of millions of people overlooked in the initial market analysis.

Case Studies: Prioritizing Inclusion Across Industries

Statistics supporting the business case for diversity and inclusion are certainly convincing, but perhaps the stronger argument is in the form of case studies illustrating wins and losses—examples of organizations that succeeded by making inclusion a priority and of organizations who suffered the consequences of not doing so. Here are a few case studies that drive home the value of diversity and inclusion in business.

Going beyond default shapes and sizes

Have you ever wondered why we had never witnessed an all-woman spacewalk until just recently? It's because the spacesuits didn't fit and only came in small, medium, and large.

When products are created without prioritizing diversity, people representing certain demographics may be left out. On its surface, not having a greater selection of sizes in spacesuits may seem trivial, but it creates a selection bias that extends beyond women to anyone not tailored for the suit. This restriction results not only in excluding certain individuals from specific space missions but also limiting the diversity of experiences and insights that could be gathered on those missions.

University of Oregon Sports Product Design Director Susan L. Sokolowski attributes this lack of inclusive design in spacesuits to a combination of budgetary constraints and underrepresentation of women in leadership positions. As Susan explains, "In my opinion, the issue of not having enough spacesuits on hand lies within what I call the product ecosystem...Most often budgetary decisions for research and development happen outside the product research and creation teams, and those decision-makers are often men. Considering representation of women at the CEO level, only 4.2 percent of all Fortune 500 CEOs are women."[2]

According to Sokolowski, who does considerable research into inclusive design, creating performance products for women requires much more than merely downsizing men's products. The most obvious differences in women are body shape and size. Product developers

---

[2]https://fortune.com/2016/06/06/women-ceos-fortune-500-2016/.

who are committed to inclusion study carefully prior to designing performance clothing for women. They have several tools at their disposal, including 3D body scanning, anthropometry, and statistical analysis of the data. The results of their research influence not only the drawing of patterns, but also the engineering and selection of materials and how technology is tailored to the body, so users can perform their jobs safely, efficiently, and as comfortably as possible.

Designer Olivia Echols considered both the unique physical and physiological characteristics of women when she designed the Nebrio Space Bra as her Apparel Studio Final Project in the winter of 2018.

Granted, in the world of apparel design, women are not an underrepresented group as a whole, but historically they have been underrepresented in industries that develop performance wear and other products for work and sports. Given that nearly 50 percent of the world's population is women, building products that fit, function, and enrich women's lives is crucial for business growth and vitality.

Thinking beyond *binary* gender is also critical, and something that is overlooked frequently. How would non-binary body shapes adjust to different designs? How can we accommodate our products, services, and (even) forms to people that identify with non-binary genders?

Designing comfortable virtual reality headsets for everyone

When Google set out to create a virtual reality (VR) headset, design team members were committed to building a headset for everyone. They were already well aware of many of the challenges they faced—people with larger or smaller heads, people who wear glasses, and so on. In addition, foreheads and cheekbones can vary by sex, by race, and more, and if you have hair with more volume, these differences can affect fit and comfort and the ease of putting on the headset.

To build an inclusive VR headset, the team tested the product, throughout the process, with people of color, people of different sexes, people with different hair textures, and more. Team members used some interesting approaches, including 3D-printed masks that tested the distribution of pressure over the face with the goal of distributing pressure evenly across each wearer's forehead and cheekbones.

During the design and development process, the team did not experience any great "aha" moments, but their efforts to include diversity in the process did influence the final product. In particular, the team chose to create the headset with a soft, flexible, breathable fabric based on the "clothes we love to wear," to optimize comfort. Extra space around the eyes was added to accommodate those with glasses. Straps were designed to enable the product to be adjusted easily for different head sizes. The final product was made with a combination of foam and fabric to make it lighter.

We now have a working group that covers all of our hardware devices. Our exec sponsor for hardware devices and services is Leslie Leland. She helped build the engineering infrastructure and processes for teams so that they could start to integrate product inclusion principles into our hardware design moving forward.

Ignoring the impact of airbags on women and children

When product design ignores certain demographics, the results can be catastrophic or even fatal, as was the case with early designs for airbags in cars. The team developing the first airbags was all male, they used a male-based height and weight chart to perform their calculations, and they tested their prototypes on crash test dummies built like the average man. As a result, women and children were killed by those airbags when they were first deployed in vehicles.

In 2006, the U.S. National Highway Traffic Safety Administration (NHTSA) amended its crash test regulations to require that airbags be tested using crash test dummies that were smaller:

> NHTSA is amending its safety standard on occupant crash protection to establish the same 56 km/h (35 mph) maximum speed for frontal barrier crash tests using belted 5th percentile adult female test dummies as we previously adopted for tests using belted 50th percentile adult male dummies. The agency is adopting this amendment to help improve crash protection for small statured occupants. The new requirement is phased-in in a manner similar to the phase-in for the 56 km/h (35 mph) maximum speed test requirement using the 50th percentile adult male dummy, but beginning 2 years later, i.e., September 1, 2009.[3]

---

[3]https://www.federalregister.gov/documents/2006/08/31/06-7225/federal-motor-vehicle-safety-standards-occupant-crash-protection.

Some of the most frequently used products have not been built for everyone, and the result can range from a mere annoyance to life-threatening or even fatal. Anything on this spectrum should be avoided if possible. Failure to integrate inclusivity into the product design and development processes also deeply restricts the opportunity for innovation—for developing game-changing solutions that improve the world and build a brighter future. Prioritizing a few key actions to promote inclusive design can lead to positive results that serve the core challenges of all your current and future users.

How Google's Product Inclusion Team Got Its Start

Integrating product inclusion into your product design and development processes does not need to be an onerous task. You can start slow and build momentum gradually, and you are likely to have a fun, challenging, and exciting journey along the way. At Google, product inclusion began as a 20 percent project—a project to which Google allows any employee to devote up to 20 percent of their time. Many of the coolest products with the greatest impact got their start as 20 percent projects.

A few of us Googlers decided that the diversity and inclusion conversation could make for an interesting 20 percent project. At the time, when we talked about diversity and inclusion, we were usually discussing culture and representation; we were not discussing these topics in the context of product or business development. However, my manager/director at the time, Chris Genteel, had spent several years looking at how small and medium businesses were less likely to be online and get found. About eight years ago, I was on a 20 percent project with Chris helping him to get these businesses online and boost their visibility. This work provided the inspiration for looking at how diversity and inclusion could improve business outcomes.

Our team began to work with other teams that were open to experimenting with us and figuring out what we were all doing along the way. A few of us, (with support from our director, Yolanda Mangolini) began to notice areas of overlap in our different work functions. At the time, Allison Munichiello and I were Diversity Business Partners—we were tasked with creating the holistic diversity and inclusion strategy for specific product areas alongside our senior-most leaders. Chris was the leader of our business inclusion work, and had Allison Bernstein on his team who had been looking at the intersections of business and the digital divide. Chris also had close ties with the advisors of several of our employee resource groups like Randy Reyes. Together, we found that all of our work functions were integral to the product design process. We knew that our communities were a key factor in getting the perspectives we needed, and we knew that senior leaders were invaluable because they set strategy, accountability, and tone.

Two roles we had not tapped previously as a team were product team leaders and members, unless they were specifically working on diversity initiatives. Allison and I, by the nature of our work, had access to product managers, engineers, technical product managers, user researchers, and more. We also had strong senior advocates like Sowmya Subramanian who helped us navigate different product areas and found teams willing to collaborate. Together, we were able to develop a glimpse of what integrating inclusion in the product design process (what eventually evolved into product inclusion) could look like.

When we thought about Google, we envisioned the company as a portfolio of products. They were all different and at different points of maturation. But we also knew that at Google, we were encouraged to think big, to thrive in ambiguity, and to solve meaty problems. We wondered how we could expand on bridging the digital divide and then internalize that solution to figure out a way to be intentional, across Google's portfolio of products, about serving users globally.

The first time we actually put our thoughts into practice was when an engineer, Peter Sherman, came to our team asking for a checklist to bring back to his team. He wanted to make sure he was thinking about diversity and inclusion in his work. Well, I didn't have a checklist, but was certainly eager to kick around ideas with Peter and his team.

Peter was amazing in that he was very open to speaking about topics that can be uncomfortable for people to talk about at work, namely race. He immediately put race on the table, saying that when working with proximity sensors and cameras, he realized his team did not have the racial diversity needed to build a truly inclusive product. (See the nearby sidebar.)

Our first product inclusion partner

—Peter Sherman, Google Engineer

Just prior to collaborating with what was to become Google's product inclusion team, I was working on a project that had a camera and a proximity sensor, both of which needed to work well for all users, regardless of skin tone. The development team was very small, though, and we had difficulty figuring out how to get the broad coverage we needed to make appropriate tuning decisions and subsequent validation testing. I knew that it was extremely important to ensure that the product was inclusive, but I didn't have the tools or the resources to adequately address the need. I was also well aware of the fact that this is a sensitive topic, especially given the challenges around racial diversity in the tech industry. Context and framing are really important when architecting a cross-sectional test plan like this, to avoid unintentionally offending participants.

At the time, there were no guides, courses, or teams internally (and it was even hard to find such things externally) to advise on best practices for approaching this potentially charged subject, and I wanted to be sure my actions were always thoughtful and sensitive. It was at this point that I learned about the diversity team and the 2015 Multicultural Summit where there would be a team and an audience eager to engage in the often difficult discussions and processes required to achieve the goal of making our products more inclusive.

During that summit, many of the participants voiced first-hand experience with cameras failing to produce pleasing images as a function of skin-tone, but few knew the history of bias in the development of color photography and that we really can improve the experience if we think and develop inclusively. For those who don't work specifically in imaging related fields, it is easy to dismiss such issues as limitations of the technology.

It was very energizing to see the excitement and engagement of the summit participants as we discussed how to work toward the

(continued)

(*continued*)

goal of de-biasing the sensor tuning process, both within our own company as well as the broader ecosystem by establishing inclusive testing guidelines and best practices. While technical challenges will always exist, development with inclusive thinking can go a long way towards creating a better, more equitable experience for all users.

We brought Peter's challenge to a summit our team ran for Googlers who represent the diversity of Google's workforce. We coupled this summit with an inclusive design summit led by Allison Bernstein for champions who were already involved in diversity-related 20 percent projects. During these summits, we heard from several leaders who had discussed the importance of bringing inclusion into their work, even if we didn't have a formal structure for what this could look like.

Googlers who were doing this work outlined a few things they needed from the diversity and inclusion team—advocacy, infrastructure, and frameworks. Because diversity and inclusion are not core to their roles and their work and because inclusion looks different for each person involved in the work, they had difficulty crystallizing the potential impact and seeing the opportunities at times.

These early discussions with different teams started to spark ideas and light a fire for our team. We began thinking through what data we would need, what services we could provide, and how we could start to create practices that could be repeatable for teams to grow and learn. We began to realize that this idea around intersectional inclusion in products could make a huge impact, so we embarked on a journey to figure it out. Our 20 percent group ultimately evolved into what is now known as *product inclusion*.

Early in 2019, our product inclusion pod (the name for a small team at Google) came up with the following mission and vision:

Google designs useful products where underrepresented users are centered, validated, and uplifted. Google also centers the voices, technical expertise, and perspectives of underrepresented communities to address product opportunities and inequities, and values their contributions as employees, partners, and users. The work is shared by everyone—all Googlers from the grassroots through leadership levels are accountable to be allies in building inclusively.

As your organization sets out on its journey to integrate inclusive design into product development, marketing, and other areas, I encourage you to focus most of your effort, at least initially, on changing minds and culture. When people in your organization fully grasp how exclusion affects consumers and how inclusive design can positively impact their lives while driving innovation and growth, they will eagerly embrace inclusive principles and practices. And as they interact with a greater diversity of coworkers and customers, they will begin to develop the empathy and understanding necessary to build products for everyone.

Chris Genteel, Director of Supplier Diversity

As the founder and leader of the business inclusion team, I've spent over a decade seeing Google work to empower Googlers and their communities to create a more inclusive culture, a more diverse Google, and equitable outcomes for everyone through our products and business.

Product Inclusion is about bringing everyone together, no matter what level they are in the company. The work is both top down and bottom up. Our inclusion champions and Employee Resource Groups drive innovation in DEI through local experimentation as well as new approaches that my team was able to harness and scale across the company for 3 years. Through these programs and more (like our Supplier Diversity program) I have led programs & partnerships that have driven $500M+ in direct economic impact for entrepreneurs from communities of color and women and LGBTQ+ entrepreneurs while driving bottom-line and user trust outcomes.

(continued)

(continued)

As a director that also focused on business inclusion, it's paramount that leaders and companies understand that product inclusion can also amplify business opportunities in addition to bridging the digital divide across multiple dimensions of diversity.

Through this process we harness the breadth of perspectives needed to create products that better reflect and serve Google's diverse users, and to create a more inclusive culture.

CHAPTER 2

Google's Capstone Research: What We Learned

Prior to embarking on any new business initiative, especially one like product inclusion that has so many unknowns, conducting research to develop a deeper understanding of the subject matter is always a good idea. On our product inclusion team, we began by experimenting and let our pilot programs guide our progress. We knew that inclusive product design is the right thing to do. However, as we got further into our work, we recognized the dearth of in-depth research in this area and recognized the need to gather and analyze additional data.

Existing research was limited in terms of looking across multiple diversity dimensions, the intersection of those dimensions, and how differences among individuals influenced both people and product. We wanted to prove the business case for inclusion and use our proof as a mechanism to help move people from awareness to action. Our goal was to prove that diversity, equity, and inclusion should be core business values and integral to product design processes. We wanted to understand how diverse perspectives impacted product design, and also how users connected with brands that demonstrate a deep commitment to diversity and inclusion.

With those objectives in mind, we drafted a research plan in the hopes of answering a few key questions and deepening our insight into inclusive design principles and practices. In this chapter, we present the questions we asked, how we set about answering them, and the information and insights we gleaned from the results of our research.

Understanding Our Research

Over the course of about nine months in 2019, our product inclusion team conducted capstone research to ensure that our diversity and inclusion practices truly add value. (*Capstone research* is a process in which students conduct an independent investigation into a question or product of their choice and produce a paper that reflects a deep understanding of the topic.) Our research could be compared to a *proof of concept*—an exercise in proving or disproving the feasibility and potential value of a proposed initiative.

We began with the hypothesis that if we apply inclusive research, design, and engineering principles throughout the entire product development process, then we will create products that perform better and are more relevant and useful for a more diverse consumer base. We were committed to designing experiments and other research activities that would challenge or support our hypothesis.

Prior to conducting the research, we carefully prepared by defining our objectives; formulating research questions; defining key terms, parameters, and objectives; and assigning teams to conduct experiments and other forms of research.

Our research objectives

Our team was already convinced of the positive impact inclusive design would have on people's lives, but we were also aware of our own bias in favor of inclusive design and the risks of not consulting people with different perspectives from our own. To overcome the inherent limitations of a small team, we consulted with a diverse group of researchers like Dr. Alva Taylor, eight executive sponsors, one 20 percent user experience (UX) lead, Giles Harrison-Conwill, an Analyst Lead, Thomas Bornheim, and many volunteers. Together, we defined our research objectives:

- To determine whether people who are underrepresented bring additional value to the product design process by way of bringing diverse perspectives to the table, leading to a richer end product. (If we found that they did, the study would provide us with data to support both a human and a business case for integrating inclusion in product design.)
- To find out which product inclusion practices we were already using were working, in what combination, and why, so that we could help both new and experienced teams understand and adopt best practices.
- To discover what team dynamics or behaviors would lead to positive results, we could work with teams and the organization overall to develop a more inclusive culture.
- To find out for ourselves why our work was important—why did we care about it, and what did we hope to accomplish through our work?
- To challenge our beliefs and assumptions in order to further our learning and mitigate bias and gaps in our understanding.

Our research questions

After discussing our objectives, we settled on the following three questions to guide the design of our research:

- Do diverse perspectives produce more successful products?
- Do inclusive product design practices lead to positive business outcomes?
- Do companies that outwardly demonstrate a commitment to inclusive design practices see an increase in engagement from underrepresented users and majority users?

Each of these questions has a well-defined purpose, ensuring that the questions aligned with our research objectives.

The purpose of the first question (*Do diverse perspectives produce more successful products?*) was to help us determine how the perspectives of underrepresented minorities (URMs) affect product outcomes. Specifically, we set out to answer the following two questions:

- Are URMs more likely than other users to call out a lack of diversity in product design?
- Are diverse teams more likely to come up with innovative ideas?

The purpose of the second question (*Do inclusive product design practices lead to positive business outcomes?*) was to help us understand more about team dynamics. We noticed that some teams integrated product inclusion from the beginning, whereas others waited until the product was nearly ready to launch. We wanted to find out precisely why some teams prioritized inclusion to the point of integrating it from the onset and identify how they practiced inclusive design on a daily basis. How did they arrive at the practices they had adopted? Could any patterns be identified across teams to integrate inclusion throughout the process? How did similarities and differences in practices affect the final product?

The purpose of the third question (*Do companies that outwardly demonstrate a commitment to inclusive design practices see an increase in engagement from underrepresented users and majority users?*) was to evaluate how users from underrepresented dimensions reacted to inclusive marketing. Because we approach product design as a holistic process encompassing every phase from ideation to marketing (and all points in between), we wanted to look specifically at the marketing component of our work. We wanted to understand how both majority and underrepresented users reacted to a public commitment to inclusion, whether their reactions were similar or distinct, and whether the commitment to inclusion reflected in the marketing would lead to greater alignment of users with that company's brand.

Definitions of key terms, parameters, and metrics

To ensure consistency across all of our research teams, we agreed on the following terminology, parameters, and metrics. As you progress through this chapter, these terms, parameters, and metrics will clarify your understanding of the data:

- **Underrepresented users** who participated in our research included the following:
  - In the U.S., Black and Latinx (a person of Latin American origin)
  - Globally, women, LGBTQ+, lower socioeconomic status, people with disabilities, age 65+, varying education levels

Note that we know this is not all encompassing, but these were the initial parameters due to the data we were able to collect.

- **Inclusive practices** are defined as key practices along with practices that come from specific operating models:
  - ○ Key practices include inclusive UX design, user testing, and team composition (representation).
  - ○ Operating models include practices developed and adopted by groups within Google, including ones that work for nascent teams and products and those for more mature products.
- **Business outcomes:** Inclusive practices were considered successful when, as a result of those practices, the business experienced a/an:
  - ○ Positive uplift in brand sentiment/loyalty
  - ○ Increase in daily active users (DAUs) with new/underrepresented demographics
- **Successful products:** Product inclusion was considered successful when the resulting product was:
  - ○ Expanding into strategic markets
  - ○ In-line with future demographic growth/changes

Desired outcomes

We recognized that research had already been done that correlates diverse teams with increased business success, but little data was available to shed light on what actions are taken to create those outcomes.

Through our research, we wanted to glean actionable insights and data that we could use to draw a connection between inclusive design practices and product and business outcomes. Specifically, we wanted to answer the following questions:

- Which inclusive practices work?
- Which methods work in part or as a whole?
- Which combinations of actions or methods work (which work better than others and in what context)?

In addition, we wanted to gather evidence to support the case for product inclusion by enabling us to do the following:

- Provide concrete case studies that illustrate best practices.
- Show what outcomes are clearly positive and clearly negative and what is going on behind the scenes with teams that practice inclusive design.
- Identify clear wins and clear failures and provide reasons for each.

Examining the data

After months of experimentation, interviews, surveys, shadowing, and simulations, we arrived at the following data summaries:

- Underrepresented consumers AND majority consumers prefer brands that demonstrate diversity, though underrepresented users are more likely to feel strongly about their product or service.
- Inclusive product design is often framed as benefiting only under-represented communities, but product inclusion shows companies are leaving opportunity on the table. Bringing inclusive perspectives can improve the overall engagement and satisfaction of those often considered majority customer targets.
- Any team that prioritizes inclusive design can build more inclusively—it is not just teams that have significant representation of women or people of color that have positive product inclusion outcomes. (Note, for the purpose of our initial research we focused on these two dimensions of diversity, but this can be expanded to all the dimensions of diversity product inclusion covers.) However, all teams that did build inclusive products consistently brought in diverse and underrepresented perspectives at multiple parts in their process.
- 100% of teams that built products that had inclusion in mind did so at multiple parts in the process (at least two key parts from the ideation-UX-user testing-marketing framework).
- Out of the teams surveyed:
  - 53.8% brought inclusion into the early ideation phases.
  - 46.1% brought into the UX phases.
  - 69.2% brought into the user testing phase.
  - 46.1% brought into the marketing phase (though product teams are not necessarily leading on the marketing front). These teams are intentionally thinking about marketing in addition to the marketing

department that does so full time and thinks about inclusive marketing guidelines and principles.

- 100% of teams that built inclusively brought product inclusion practices into at least two of the four phases (ideation–UX–user testing–marketing).

Drawing conclusions

After examining the data, we came to following conclusions:

- Overall, inclusive design practices have a strong positive impact on how users perceive and engage with a product. This is true across all age groups and for both majority and underrepresented users.
- Building for underrepresented users actually is not only creating a social benefit for everyone, but now we know it also is a business benefit.
- Both underrepresented and majority groups prefer inclusive products.
- Firsthand perspectives are crucial; theoretical empathy is not enough. As Google Director Matt Waddell advises, "Love your customers and love their problems, and show up however you can."
- Building for everyone requires being intentional about addressing the needs and preferences of historically underrepresented consumers; otherwise, underrepresented groups will be overlooked.
- Gathering data and tracking metrics are keys to success. Product inclusion must be treated as a business function and priority. See Chapter 11 for more about measuring product inclusion performance.
- Teams that have less diverse representation can also create truly inclusive products if they set intentional goals to do so and prioritize speaking to underrepresented users at key phases in the process.
- Managers of teams that build inclusively note that bringing inclusion in early is the key to success. Inclusion isn't about checking a box, but integrating inclusive design early and through rounds of iteration.

Having concrete data that supports the business case for inclusive design is very exciting; it proves that product inclusion results in better products and services and shows that bringing diverse perspectives to the table leads to increased innovation.

Learning from Our Experience

After years of experimentation, iteration, and trial and error, here are some highlights of what our team has learned in regard to product inclusion:

- Diverse perspectives lead to increased innovation and products that are better not only for underrepresented users but also for majority users.
- Ideation, user experience research, user testing, and marketing are core areas to focus on, bringing an inclusive lens to the process, even when each area commits to a single change or action.
- Product inclusion should be embedded into processes, not a stand-alone idea or a process tacked on at the end.
- Applying an inclusive lens to product design and development does not necessarily result in delays. It involves designing more intentionally.
- Billions of users are yearning to be seen in products and have the purchasing power to act when they've been brought into the fold.
- The further from the "default" user a consumer is, the more likely that person will feel alienated from a product or service.
- Product and marketing teams need to be intentional (proactive) about identifying and addressing the needs and preferences of historically underrepresented consumers. Product inclusion doesn't just happen.
- Diversity, equity, and inclusion (DEI) energizes members of an organization as soon as they understand its purpose and importance and have concrete, actionable ways to integrate it with their work. Don't assume that people will be reluctant to embrace product inclusion initiatives. Once people understand the reasoning, from a consumer and business perspective, they will eagerly embrace it.
- Organizations do well by doing good. Building products and services that appeal to more consumers will undoubtedly grow any business.

Putting Our Research and Experience to Work in Your Organization

I encourage you to put our research to work in your organization to increase awareness of underrepresented users, their needs and preferences, and the business opportunities they represent and to start integrating inclusive design principles and practices into all stages of the design process from ideation to marketing. Here are several ways to integrate product inclusion into a team or across your entire organization:

- When planning for your next quarter or year, include diversity and inclusion as part of your plan. Budgeting time and money for inclusive design training and initiatives.
- Bring more diverse perspectives to the product design and development process by making teams more diverse, recruiting URMs internally and externally to participate in the process, and talking with users from underrepresented communities.
- Integrate product inclusion throughout all four stages of the design process—ideation, UX research and design, user testing, and marketing (covered in Chapters 7 through 10).
- Identify the critical points in your design process and create commitments around them. (See Chapter 6 for guidance.)
- Prioritize product design for URMs. Unless you make a conscious, intentional effort to design for URMs, they will be deprioritized and left out of the process altogether.

The world is celebrating and demanding diversity and inclusion, not just in culture, but in products and services. To learn more about how to make your organization more diverse and inclusive, visit accelerate .withgoogle.com.

CHAPTER 3

20 Essential Product Inclusion Questions to Light the Way

In many ways, product inclusion is one big learning process. When you are doing it right, you are continuously learning about groups that are underrepresented, how to build more diverse and inclusive teams, how to integrate diverse perspectives into your product design and development process, how to access resources to support your efforts, and much more.

This learning process is most productive when you and your team have a natural curiosity and ask the right questions. When you are getting started, however, you may not know enough to ask the questions that elicit the answers and guidance required.

In this chapter, I present two sets of 10 questions—10 questions every team needs to ask itself (and answer) and 10 questions our diversity and inclusion team is commonly asked (and our answers). Along the way, I explain the significance of each question, and I answer each question or provide guidance on how to answer it yourself or with the assistance of other team members.

The 10 Questions Every Team Needs to Ask and Answer

Approximately one year into Google's product inclusion journey, questions were streaming in from across the company at a pace faster than

I could answer them. To meet everyone's needs while fulfilling my other responsibilities, I decided to hold office hours; I set aside one hour of every day to meet with fellow Googlers about product inclusion. People could sign up for 20-minute appointment slots to ask questions, obtain guidance, and provide input on team-based or Google-wide product inclusion topics.

To optimize the productivity of these meetings, I require that all participants (including me) be prepared. Several days prior to each scheduled meeting, I send the future participants a list of 10 questions, which our team came up with, to find out their current level of understanding (beginner, intermediate, or advanced), if they have a product inclusion strategy in place, the nature of their work, and who their partners or stakeholders are. These are the questions (covered in greater detail in the sections that follow):

- Has your team been exposed to product inclusion?
- Have you identified a champion for your product inclusion efforts?
- What is the product (or business) challenge you're trying to solve?
- What is the inclusion challenge you're trying to solve?
- How do the product/business and inclusion challenges align?
- Whom do you need to influence to unlock resources to solve the problem?
- What's your action plan for a test/pilot?
- What partners need to be involved to execute, document, measure, and communicate the results of your test/pilot?
- How can you build the resources to continue this work beyond this workshop?
- What is your public commitment to documenting and sharing the outcomes of your work in product inclusion (both internally and externally)?

This preliminary Q&A process helps us manage our bandwidth, prioritize work, and ensure that teams come back to share information and insights about the challenges they encountered and the changes they made, so we can continue to learn and to gather knowledge and insights to share across the company. This process also ensures that teams perform the background work needed (identifying a champion, getting clarity on their current process, and writing down any ideas they have) to optimize the productivity of these consultations.

I approve a time slot only when I have the following items from the person requesting the meeting:

- A clear agenda
- Clear reasons why each person should be in the meeting
- Reasons why each person has an interest in product inclusion
- A commitment that participants would share their learning with our team
- Assurance that the participants had already followed instructions on how to get started on our product inclusion team site

In addition to optimizing the productivity of meetings, preparation for meetings ensures that each team has people committed to implementing what is required to make product inclusion a success. As with any part of the product development and rollout process—from deciding what to build to building and testing to marketing and sales—multiple people will be involved with signing off and giving feedback. Product inclusion works best when a team has clear owners and the product inclusion specialist serves as advisor. If you are just starting out, you may not have someone who qualifies as an "expert," but you should have one or more individuals who can serve in this role—people who have stepped up to help the organization learn, grow, and prioritize product inclusion.

The questions we ask prior to consulting with teams can serve as conversationstarters among team members or as a tool for champions to lead the initiative. Depending on your role, the makeup of your team, and the nature of the work, some of the questions may be more or less relevant to you and your team and the work you do, but these questions and their answers provide a framework from which to start.

Note that the questions are a mix of highlevel questions (What are you trying to solve?) and tactical/personnel questions (Who is going to help you solve them?). They can be answered by the point person/champion who is helping to move the work forward or by the team, if members are already committed. However, one or more people responsible for finalizing the answers should have a clear understanding of the team's objectives and strategies and a general idea of how an inclusive lens is likely to add value to the work. A person with this understanding is better equipped to answer the questions in the context of the team's mission.

One last word of advice before I dig deeper into these questions: do not be concerned if you cannot answer all of the questions. The process of asking these questions, answering them, and sharing and discussing your thoughts with others will bring clarity over time. Questions 2 through 7 in particular help to clarify what you are trying to solve, how inclusion aligns with your current work, and who can help your team integrate product inclusion. The answers to these questions are key to moving forward.

Has your team been exposed to product inclusion?

Knowing how much your team knows about product inclusion is important in determining your starting point. A team that is unfamiliar with product inclusion may need to start with a general understanding of what it is and why it is important, which is covered in Chapter 1 and throughout this book. On the other hand, a team which is already well-schooled on product inclusion and eager to get started is likely to benefit more from guidance or training on higher level or on issue-specific topics, such as choosing or developing tools, or best practices for testing products and obtaining feedback. Sometimes, all an experienced team really needs is a referral to another team that has already encountered a similar product inclusion challenge and has overcome it or may have a resource that can help them achieve their goal.

Although this is worded as a yes/no question, I strongly encourage you to provide an in-depth answer that reflects your and your team's understanding of product inclusion and perhaps even exposes gaps in your knowledge and expertise. In other words, try to look at what you and your team already know and what you and your team need to know or would like to learn to improve your processes, practices, and outcomes. Identifying a gap in knowledge (seeing what's *not there*) is not always possible, but teams are often aware of what they don't know and need to learn.

Have you identified a champion for your product inclusion efforts?

A *champion* is an individual who takes a genuine interest in the adoption or implementation of an initiative and will voluntarily invest time, effort, and resources in promoting it. Champions can be at any level in an organization, but generally the higher up in the organization, the more influential that person is and the greater access the person has to resources.

If your team is planning to integrate product inclusion into their work, they will have a better chance of success by rallying the support of a champion. Try to identify that influential someone in your organization who is willing and able to support your team's efforts or someone who may at least be receptive to the concept—someone you can recruit to become your ally. (See Chapter 4 for guidance on how to recruit people to join your cause.)

If you have a champion in mind, answer this question not only with "yes" or "no" but also with the person's name and position and relevant additional details—interests, responsibilities, level of product inclusion understanding, and so on. The more you know about your champion, the better equipped you are to communicate about product inclusion in a way that resonates with that person.

What's the product (or business) challenge you're trying to solve?

When you ask this question, your focus shifts from how to get started (your point of departure) to your destination, objective, or desired outcome. Thinking about the end may seem odd given that you are trying to figure out how to get started with product inclusion, but having the end in mind is important when you begin to formulate any plan. When you have a clear idea of what you want to accomplish, you are better prepared to start figuring out which pieces need to be in place and the direction and steps to take.

For now, think about the product or business challenge you are trying to overcome. For example, you may see that people over 65 don't engage with your product and bounce from your website within the first minute. You'd like to increase your market share with this group by 15 percent as you've had only one percent growth in revenue in the past year.

If you are uncertain about the challenge you are trying to solve, see Chapter 7 to find out how to research underrepresented users. Research often reveals problems or challenges that open the doors to innovation and opportunities.

What is the inclusion challenge you're trying to solve?

The inclusion challenge shifts your focus from business or product development to people. Who are the underrepresented individuals you want your product to serve? What demographic(s) or intersection of demographics are you and your team seeking to include in the product design and development process? What is it about products or services currently being offered that is falling short of meeting the needs and preferences of certain demographics?

How do the product/business and inclusion challenges align?

After answering the previous two questions, you have some idea of the product/business challenge and the inclusion challenge that you and your team are trying to solve. Examine the two challenges closely to determine how they are aligned or where they intersect or overlap. For example, if you realized that people over 65 were bouncing from your site and not purchasing products, and you had not tested with anyone over 65, you may not realize that the site has an inclusion issue (for example, the font is too small and the text color too light, or the branding does not resonate). Note that addressing this challenge would benefit more than just the initial target demographic; it also would benefit people who wear glasses or have more severe low vision, for example.

Aligning your product/business challenge with your inclusion challenge amplifies the impact of your product inclusion efforts by ensuring that the human case for inclusion aligns with the organization's business strategy and objectives.

Whom do you need to influence to unlock resources to solve the problem?

Decisions about how to allocate resources may significantly impact efforts to implement product inclusion. To improve your chances of success, determine the resources you need and identify the person(s) who hold the keys to unlocking those resources. Resources may include the following:

- Personnel/expertise, which may include sharing across teams
- Time allotment for inclusion volunteers to participate in meetings and testing
- Money for anything that needs to be purchased, such as special materials or support, or costs of travel for research
- Space to do user research or run design sprints
- Access to certain tools/data to get to the root of the challenges you are trying to solve

When you have a comprehensive list of resources, identify one or more decision-makers in your organization who can provide access to those resources. Ideally, you can find a single individual (an executive or supervisor) who is willing to take on the task of procuring resources.

What's your action plan for a test/pilot?

Having a plan in place, whether that plan is clear and detailed or general and tentative, is important both for implementing your initiative and for recruiting others to join your cause. If you approach a supervisor or executive with your idea for developing an inclusive product, you will be much more convincing if you have a viable plan in place.

If you are having trouble formulating a plan, meet with team members and anyone else in your organization who may be able to provide the input and guidance required to draft a plan.

What partners need to be involved to execute, document, measure, and communicate the results of your test/pilot?

Product inclusion is a team sport, a collaborative effort. You and your teammates may be able to do much of the work required, but you are likely to need some outside assistance; for example, people in the organization who have the expertise you need and people who represent the population for which you are building the product. You may also want to partner with your organization's communications team to document and share your work—the challenges you encountered, your successes and failures, and so on—for future learning.

Make a list of the partners who can contribute to your efforts; for example:

- **Product managers** are integral to get on board because they oversee the end-to-end process.
- **Marketing team members** are great to partner with because you need people who can drive the external work needed to tell authentic and inclusive stories.
- **User experience (UX) researchers and designers** help run inclusive research and to build plans for product inclusion.
- **Executive sponsor(s)** can help build buy-in and implement accountability frameworks to get the organization to take action.
- **Analysts** can help identify opportunities, assess progress, and create cohesive ways of pulling the data together to present to key stakeholders.
- **Diversity, equity, and inclusion teams or individuals** can build product inclusion into what they do to help amplify the work in a strategic way.

How can you build the resources to continue this work beyond this workshop?

To answer this question, think in terms of "What's next?" Create a list of concrete actions your team can take right now to start moving toward your desired outcome. Here are a few examples:

- Our team will do inclusive testing for 30 percent of our products this year, and increase to 50 percent of products next year.
- Our team will do one international research trip a year.

- Our team will partner with employee resource groups (ERGs) to understand the top product challenges and opportunities from their communities. (ERGs are employee-initiated groups, financially supported by Google, that represent social, cultural, or other underrepresented communities.)

What is your public commitment to documenting and sharing the outcomes of your work in product inclusion (both internally and externally)?

Our product inclusion team always poses this question to individuals or teams who are seeking our assistance, because we want to ensure that we can gather their feedback to further our own learning and to share what we learn from their experience with others across the company and with external organizations.

Sharing internally signals that your organization is committed for the long term and is serious about building inclusively. Product inclusion should be treated as any other core product priority and should be documented and measured. See Chapter 6 for more about structuring and documenting product inclusion efforts and Chapter 11 for details about measuring progress and performance.

Sharing externally signals to existing and new users that you care about building for everyone, with everyone. Both underrepresented and majority consumers and users are more likely to feel an affinity to your brand if you demonstrate a commitment to product inclusion. This fact is supported by data from research we conducted. (See Chapter 2 for details.)

10 Questions Most Teams Ask Our Product Inclusion Team

After every product inclusion presentation I deliver, I open the floor to questions. A frequent question asked is "What do you wish teams knew in the beginning of their product inclusion journey?" This is a difficult question to answer because teams need to know quite a bit at the beginning of their journey, but by answering the 10 questions covered in the previous sections, teams will be well prepared to start the process—regardless of their industry or the nature of their work.

Following are 10 additional questions I am often asked after delivering a presentation, which I answer in the sections that follow:

- Why should leaders care about team members' personal backgrounds?
- Why is culture important?
- What part of product design is the most important?
- Do you have insights specifically for [insert specific product/service]?
- Can you come to my team meeting?
- Which dimensions do I prioritize?
- How do I get the most perspectives to the table?
- I want my program manager/VP/marketing manager/engineering lead to care about this. What do I do?
- Am I biased?
- What's the one thing I should focus on?

These questions and their answers lead to valuable information and insight that can help teams that are just getting started with product inclusion. And if you are the resident product inclusion specialist at your organization, these questions and answers can help prepare you for some of the most common questions you are likely to be asked.

Why should leaders care about team members' personal backgrounds?

At some organizations, the culture discourages the sharing of personal experiences, activities outside work, personal hopes and dreams, and anything else that makes a person unique. A division exists between home and work, which may carry certain benefits; for example, employees may be more focused at work and be less burdened by work-related concerns at home.

However, managers and executives are wise to get to know the people in the organization on a more personal level and even encourage them to bring to their work what makes each of them unique. When you demonstrate that you genuinely care about and are interested in the people who work for you, they become more likely to share their experiences and insights. As a result, people in the organization begin to take their work personally and invest a part of themselves in the process, enriching the process and giving it a stronger sense of purpose.

Research shows that diversity drives innovation,[1] and the best results are achieved when people feel comfortable being themselves. Everyone on the team needs to be accepted and included in order to feel safe enough to be themselves, to share their unique experiences and perspectives, and to challenge long-held beliefs and practices.

I remember sharing my personal experiences with a senior leader after he made the effort to ask questions. He had also invested considerable time outside work to research and understand the history of race in America and find ways to support underrepresented talent in the organization. Because I knew he cared, I was willing to open up and share, which I had felt uncomfortable doing in the past. People who are underrepresented are often concerned about sharing too much of their backgrounds or are trying so hard to fit in that they do not want to call attention to their differences. As an executive or supervisor, you want people to share and to bring different experiences and perspectives to their work.

Why is culture important?

"Culture" is an interesting word. In sociology, it is defined as the language, beliefs, behaviors, arts, institutions, and achievements of a nation, people, or other social group. In biology, "culture" refers to a medium that supports life and growth. In the context of product inclusion, I like to think of an organization's culture as reflecting both definitions. Product inclusion delivers the best outcomes when diverse members of a team (and any participants included from outside the team) engage in a shared mission and work in an environment in which every individual is encouraged and allowed the freedom of self-expression.

---

[1]https://hbr.org/2013/12/how-diversity-can-drive-innovation.

To build a culture conducive to product inclusion, focus on two key factors:

- Representation (diversity among team members)
- Environment (a culture of inclusion)

In terms of representation, two cultures require consideration—the internal (team) culture and the external (users) culture. Ideally, the internal culture is a mirror image of the external culture; the two communities reflect one another in respect to differences in race, sex, gender identity, ability, and so on. However, because the population of the outside world is far larger and more diverse than that of any organization, a team's representation (population) rarely reflects the greater diversity present in the world at large, especially in small organizations.

To overcome the limitations of representation, look for ways to bring in diverse perspectives. You have a few options, including the following:

- Pull in other perspectives from across the organization to share their experiences and insights and include them in the product design and development processes
- Converse with current users, customers, or clients from different backgrounds
- Engage with people in underserved and underrepresented communities—people who speak another language, live in another part of the world, are older or younger, have different abilities, and so on

As you bring more underrepresented people and perspectives into the process, be sure to foster an environment of acceptance and inclusion. Do not merely invite them to share; make them feel welcome and include them in the process. With a safe and diverse internal culture, people will be more likely to call out a lack of diversity and share ideas that challenge the default or majority thinking. As a result, the circle of innovation expands and encompasses the differences among a broader user base.

I once worked with a leader who repeatedly asked, "What is the worst thing that could happen if we move forward with the original plan?" This question allowed people to dissent in a safe way and to brainstorm alternative plans.

What part of product design is the most important?

At Google (a tech company), we have broken down the priority areas of the product design and development process into the following four phases (see Chapter 6 for details):

- Ideation
- User experience (UX)
- User testing
- Marketing

All four are important, but remember, the earlier product inclusion is introduced into the process, the greater the impact. If you wait until the user testing phase, you'll waste a lot of time fixing problems that could have been prevented in the ideation phase. If you wait until the marketing phase, you end up trying to sell a product that is fundamentally flawed in terms of product inclusion.

What is common among all four phases and, therefore, has the greatest impact overall is engaging with a diversity of users. Your success hinges on your ability to find users who are not like you and your team and on maintaining regular, ongoing dialog with members of underrepresented communities.

When searching for an area to focus on (an area with the greatest potential), look for any area that strikes you as low-hanging fruit—something you can act on immediately and with relative ease. For example, if you already do UX research, consider expanding that research to include a different part of the country or partnering with a local college to increase the diversity of perspectives in relation to age, gender identity, race, and ethnicity. If you already do focus groups, consider making these groups more inclusive of different gender identities or abilities. Being aware of your current process and modifying it to make it more inclusive often is the best way to start building momentum.

See Chapter 6 for more suggestions on how to integrate product inclusion into the four phases of product design and development.

Do you have insights specifically for [insert specific product/service]?

When we consult with various teams, they invariably ask if we have product- or service-specific data. We often do, but it's also important to think about the why behind needing data and how that can power insights that help your users.

I encourage you to take the same approach toward collecting and analyzing insights from users and sharing the findings with your organization. You cannot assume you know what a certain demographic needs or prefers without doing your homework and talking with people in your target demographics. Gather data on customer satisfaction, user pain points, and the different ways your product is being used (you might be surprised!). (See Chapter 11 for details about the metrics that matter.)

Two summers ago, I had a fantastic intern, Kourtney Smith, who created an internal site that shared powerful insights across different dimensions of diversity. For each group, she aligned trends in behavior, interests, and more, so that Googlers can learn more about each group and be encouraged to do their own research. Teams can submit data they'd like to share with other teams to make it available on the site. This approach not only creates a one-stop shop, but also enables teams to share their data, building community and helping contributors feel that their product inclusion work is being recognized. Data sharing reduces the learning curve and helps teams avoid pitfalls that other teams may have encountered. In a subtle way, sharing diversity and inclusion data recruits other individuals and teams to integrate product inclusion in their work.

Your team does not have to create a high-tech intranet site or business intelligence (BI) dashboard to reap these benefits—in a smaller organization, you can create a shared folder on the network where people can upload their data, or you can circulate a folder via email.

The key is to democratize the data to encourage and empower everyone in the organization to view their work through the lens of diversity and inclusion.

Can you come to my team meeting?

If you are the resident expert on product inclusion, you are likely to experience increasing demand for your time as interest in product inclusion grows. If you serve a large organization, you are likely to receive more invitations than you have the bandwidth to attend, so when someone asks, "Can you come to my team meeting?" The best answer may be, "That depends." When demand exceeds your ability to supply information and guidance, prioritize.

At first, our team would go to any and every team meeting. Now, the sheer volume of requests makes that impossible. I have to prioritize, which I do by asking myself, the following questions:

- Are they willing to act on the feedback, even if it makes them change the direction of the product?
- Does the team have a champion who can have initial conversations with team members?
- Are there opportunities to speak to a broader swath of the team; for example, offsite or all-hands team meetings?
- What is the impact to the user?
- How many users could this business, product, or functionality possibly touch?
- Does the need for improvement of this product/functionality have a history as it relates to underrepresented communities?

Ask these same questions whenever the demand for your expertise exceeds your ability to meet that demand. The answers will help you prioritize your work.

Alternatively, you can start with a part of your company that shows excitement or commitment to learning or start with the area that holds the most potential for integrating product inclusion. Maybe you have a product manager who is looking to increase opportunity for a product launching at the end of the year. Talking to a motivated leader about product inclusion may be a great start because that person owns the process and can partner with others (engineering, for example) who are likely to play an integral role in getting product inclusion "in the water."

Which dimensions do I prioritize?

Our team's mission is to build with everyone, for everyone, and setting out with that goal in mind is always best. If you can accommodate everyone's needs and preferences, your potential user base is as large as it can possibly be. However, sometimes the best approach is to prioritize by identifying the most serious unmet user needs. We avoid stack ranking dimensions, because people are multifaceted and many times, solving for one group will help many groups, both underrepresented and majority.

Although my team is often asked which dimension the team should prioritize, that is often a question we cannot answer for several reasons:

- Choosing only one dimension of diversity is not advised. You want to consider targeting several underserved communities and the intersections of those communities.
- Deciding which dimension(s) to focus on is a management or team decision and, like all business decisions, should be based on data. In other words, specific research is usually needed to answer this question.
- Every team needs to engage with underrepresented users prior to deciding which products to build and the functionality to build into each product.

For example, suppose you discover that your product skews to people in Europe, ages 50–65. You would prioritize your work on this dimension. A good first step would be to interview members of this demographic and people who are older and younger to gather information about their experience with your product and similar products, what works, what doesn't, what frustrates them, and so on.

Keep in mind that people are not defined by a single dimension; diversity dimensions can intersect and overlap. However, you may want to begin with a dimension that has not been a priority in the past, such as socioeconomic status. For example, if your company has traditionally created high-end products costing in excess of $1,000, you may want to consider researching the potential impact of a product at a lower price point with different users. Finding out the price point these users would be open to paying, what their usage with other products looks like, and the functionality important to them could be a good start. For example, suppose they use their phones primarily to access the Internet, as opposed to making and taking calls. You could prioritize Wi-Fi and Internet features while making the phone more economical.

Another example comes from my own personal experience: I am Black, a woman, and left handed. All these dimensions may affect how I interact with a certain product, but if you're designing scissors, one dimension is clearly most important—left-handedness. When you are left-handed, even a task as simple as cutting a piece of paper can be difficult with a pair of scissors designed exclusively for left-handed people. The fact that I am a woman or Black, has little, if anything, to do with designing scissors. It does come into play, however, if you are designing an app that allows people to try on different cosmetics.

How do I get the most perspectives to the table?

Even if a team has diverse representation, it can always benefit from additional perspectives. Unfortunately, expanding a team's circle of influence is often a big challenge. Here are a few ideas for bringing more perspectives to the table (see Chapters 7–10 for details):

- Create a product inclusion volunteer group and recruit people from underrepresented communities.
- Incentivize people to do surveys.
- Form a group of trusted testers—friends and family of your employees.
- Conduct online surveys or in-person surveys at highly populated locations (malls, for example).
- If you're a smaller company, bring in people from different areas of the company; for example, invite people from marketing to participate in a product meeting or create a focus group of friends and family members of employees.

Be thoughtful about bringing underrepresented users to the table in a way that is both respectful and safe enough for them to be candid about what works and doesn't work with your product. Do not assume someone wants to speak about their experiences of being underrepresented, and be sure to express your appreciation when they choose to share their thoughts.

Avoid the common mistake of bringing in people from different backgrounds merely for the sake of doing so. You can start to tokenize people in this way, making them feel that they have been brought in only because they represent a particular demographic. Being the "only" and having to represent one's culture or community is awkward and exhausting.

I want my program manager/VP/marketing manager/ engineering lead to care about this. What do I do?

One of the biggest challenges to integrating inclusion into product design and development is convincing others of its value. The best approach is to make your case—a human case and a business case, as explained in Chapter 4. You want to show the positive impact of inclusion on underrepresented users and the potential opportunity in terms of innovation, revenue, and growth.

Focus on the core challenge you are trying to solve. If your product is meant to ease the lives of people and empower them, gather statistics on those not being served and combine it with data on the purchasing power/opportunity these demographics represent. Telling a compelling story backed up by data is a surefire way to convince others.

Another way to bring people into the fold is to have them come face to face with users who may be affected by their decisions. "Getting proximate" as discussed in the introduction, can help people empathize and truly understand how building for underrepresented consumers can impact their lives in positive ways. I remember visiting a university and having the students talk me through how they chose their laptops for school. Price was a huge consideration, as were speakers to play music! Because I haven't been in college for almost 10 years, I would not have been able to put myself totally in their shoes until I heard their stories.

You don't have to be a part of an underrepresented group to get product inclusion off the ground. In fact, many of the most passionate and active champions and senior sponsors are members of the majority population.

Am I biased?

The short answer to this question is "yes"; we are all biased. The solution to overcoming bias is to recognize that fact and to encourage ourselves and one another to work proactively to mitigate our biases. Bias is not always bad; we wouldn't be able to function very well without taking some mental shortcuts. However, we need to be cognizant of that propensity and challenge it so we can become intentionally inclusive. When we take the time to ask, "Who else?" or to get feedback from someone unlike us, we are able to look at our work from a different perspective and understand how to make our end product more inclusive.

As I was writing this book, my editor would highlight areas where I needed to go deeper or expand on a concept. Because I work, eat, drink, and sleep product inclusion, I understand the concepts and what is required to integrate inclusion into product design and development, so I would unintentionally omit essential details. My editor served as a readers' advocate, bringing an outside perspective to the manuscript, highlighting areas where I needed to provide additional explanation and detail. Her input was invaluable in helping to call attention to what I was overlooking. In the same way, teams need outside perspectives to overcome their bias and shed light on issues that might otherwise be missed.

Intentional design

—*Emmanuel Matthews, Group Technical Program Manager, Artificial Intelligence at Course Hero and Founding Principal at Amplify Genius*

Design should be done with intention—you rarely achieve what you do not explicitly optimize for. When you're building a product that is intended to be used by people from all walks of life, you need to be sure that you clearly understand how your product will be perceived, how it will be utilized, and how it will impact those users. The reality, particularly when it comes to technology companies, is that your product team is likely to be a relatively homogenous group. This is based on many different factors from hiring and company culture, to historic inequities and geographic location, among others.

(continued)

(continued)

However, the impact a team's homogeneity has on a product is the same—what was missing in the process inevitably shows up in the product. These oversights can manifest themselves in a wide variety of ways, ranging from sensors that struggle to recognize dark skin to digital personal assistants that make recommendations of no relevance to the user.

Long gone are the days where global products could be designed in a way that fails to meet even a minimum bar for product inclusion. With the proliferation of social media, products that exclude users from historically underrepresented communities face a whirlwind of negative media and social backlash. Companies risk permanent brand damage, resulting in not just a loss of revenue, but also an impaired ability to recruit and retain talent from the underrepresented communities they so desperately need to help prevent these fiascos to begin with.

What makes product inclusion difficult is that it requires product owners and leaders to be honest and clear about who they are genuinely aiming to serve. Most often, leaders shy away from this admission because of the dissonance that results from building a product solely for an affluent customer base while the company's mission states that it is building to impact the world at large. The truth is, however, that building an inclusive product is not just a moral or ethical imperative, it legitimately improves the usability and function of your product in every way, shape, and form. When I build products, I intentionally design them not just for users who support our initial use case, but I imagine how users from even the most marginalized communities could benefit from them and ensure that same mindfulness permeates the development process.

Inclusion doesn't require a degree in quantum physics; it simply requires mindfulness and intention. By far, the easiest way to build an inclusive product is to get your collaborators to help. Since the people who've been traditionally excluded are probably absent from the meeting rooms where decisions are being made; it's essential to foster an environment that encourages and supports all employees who speak up about these issues. On many occasions, I reached out to Annie and our Product Inclusion team at Google when I worked there to request their assistance for things that ranged from data collection strategies to product management advice to help me cover anything my team was overlooking.

If your company doesn't have a product inclusion team, you aren't out of luck; you should simply raise questions about the inclusivity of your product consistently and add checkpoints to your roadmap where you can evaluate your product with a diverse set of users. Another easy way to ensure this happens consistently is to include a section for product inclusion in all requirements documentation—whether it's business, research, product, privacy, or quality assurance.

What's the one thing I should focus on?

The answer to this question varies on numerous factors, including your and your team's level of understanding of product inclusion in general and more specifically of the underrepresented groups you are wanting to serve, your team's composition, your existing product design process, your access to the expertise and resources you need, the nature of the product or service you want to design, and so on.

In general, the first thing all teams should focus on is building awareness and understanding of underrepresented groups, and that involves meeting and conversing with members of those groups. Only when you "get proximate" with the people you are designing for (and ideally with) can you begin to make well-informed decisions about which products to build, what functionality to include, and how to test and market your products.

Highlight demographics you may not actively be prioritizing, and then start learning from people who fit into those various demographics by talking to them, interviewing them, doing focus groups, conducting research, putting that research into the product design, and allowing them to test the product and provide feedback.

Making the world more accessible with Live Transcribe

—Brian Kemler, Product Manager for Android, and Christopher Patnoe, Head of Accessibility Programs

Live Transcribe is a communication tool to make it a little easier for folks with hearing loss or deafness to communicate with others. It was inspired by Dimitri Kanevsky, a Googler who faced a challenge communicating with a colleague. In solving this challenge for Dimitri, we realized we were on to something that could benefit many users. Live Transcribe performs real-time transcription of speech and sound to on-screen text, so users can more easily participate in conversations. As a company, "Our mission of making the world's information more universally accessible and useful" is one step closer to fruition when someone is better able to communicate with a loved one. Features are helpful and meaningful when they help empower people with disabilities to live their lives to the fullest.

On the product and process side, success looks like the completion of a culture change where people treat accessibility as if we were building for our future selves; you may not be not building for yourself as you are today, you are building for the "you" in the future (if you're lucky enough to get there) or for those you will never know.

Getting feedback from actual users rather than making assumptions on their behalf is a prerequisite to an inclusive and accessible product development. This requires user-input early and often. It means sometimes challenging our own assumptions and building bridges to new communities.

Live Transcribe shows that building for a specific group or even one person's challenge can lead to positive results and benefits for more than just the target audience. We envisioned a world where captions were available everywhere (for media on devices and for real world conversations), and through collaboration with Gallaudet, we were able to launch Live Transcribe for all users.

The four key pillars of designing accessible products

—Jen Kozenski Devins, UX Lead for the Accessibility Team

We focus on efforts that will have the broadest impact. For example, knowing that design systems are the foundation of many products, we believe that focusing on making those accessible will help make the design and development process easier for teams so they can focus on the more challenging and interesting design problems.

To fully understand where we should focus our efforts, we also conduct internal research to understand what teams need to do to make more accessible products as well as external research to better understand unmet user needs.

We focus on four key pillars to support product design:

- Provide resources, research insights, and tools to make creating accessible products easier for all teams
- Help teams understand the quality of the user experience of their product through user testing
- Support the design and development of innovations in the accessibility space (products/features design from the start to meet the needs of people with disabilities)
- Help to change the culture of UX to focus more on inclusive design and accessibility through training, external outreach, and getting into school curricula

Putting product inclusion mechanisms in place

—Nina Stille, Lead Diversity Business Partner, Intel

People remember products and experiences. Product inclusion opens new consumer markets and has the power to increase brand recognition and attract top technical talent.

(continued)

(continued)

Technical leaders take quickly to the concept of product inclusion because it's logical. As an industry, we've done a good job setting clear criteria and measurement for accessibility (but of course there is room to improve). Our challenge for 2019 was to apply learnings from the accessibility space to other dimensions of diversity, including race, ethnicity, gender, age, geographical location, etc. Senior leadership is resourcing and leading efforts to create a set of product guidelines to help product managers, researchers, and UX designers determine whether their products are inclusive. The guidelines include leveraging diverse training data for machine learning and conducting user research across diverse populations. Our hope is to take the proven concepts from the guidelines and create and implement a rating system that can measure the level of inclusion of a software product.

Our diversity councils regularly publish progress on our goals to the organization, which naturally helps to spark a good debate. I can always tell when the accountability mechanisms are working when I start to get a bunch of emails from senior stakeholders about progress or lack thereof.

CHAPTER 4

Building the Case for Product Inclusion and Getting Buy-In

To succeed in developing a product or service with universal appeal, you must first get several levels in your organization onboard, from leadership down to the people who design, develop, test, and market the product. Ideally, you would have universal buy-in on the importance of product inclusion from those responsible for each product's success. Through universal buy-in, you significantly reduce the likelihood that key actors will merely follow the status quo, performing their jobs as they always have done, while at the same time you energize everyone involved to create truly inclusive products.

To change the way people in your organization think and act, you need to rally the troops, which requires building a strong case for product inclusion and getting buy-in from top to bottom.

Building the Case for Product Inclusion

Building the case for product inclusion actually requires that you build two cases to convince others of its importance:

- **Human case:** Typically a story that illustrates the importance of product inclusion to historically underrepresented consumers.

- **Business case:** Facts and figures that present the benefits of product inclusion from a business perspective.

Of course, you may get people to come on board by using only one case or the other. If an untapped market has a ton of potential, for example, a business case may be sufficient to bring everyone onboard. However, if the business case is relatively complicated, the human case can tip the balance in favor of product inclusion by, for example, highlighting how the lives of underserved or poorly served consumers will be transformed through inclusion.

Combining data (your business case) and relatable stories (your human case) provides rational and emotional incentive for people within your organization to help you move the work forward.

Human need and business opportunity

—John Maeda EVP/Chief Experience Officer, Publicis Sapient Maeda Studio.com

Caring about inclusion is different from seeing it as an opportunity. Product inclusion is about both. You want to transition people from seeing the need for inclusion, to understanding the opportunity, and then to acting toward changes.

What makes this last step difficult? For large companies, it can be like turning a big ship. The company already has momentum, resources, and power going in one direction, and it takes time and effort to shift them. Smaller companies, on the other hand, may have limited resources, time, personnel, etc. But regardless of whether a company is big or small, empathy is always possible. Understanding someone else's perspective unlocks the opportunity to see your product or service in a different light and bring a different lens to the product design process.

For leaders, taking that first visible step can be the most difficult. Inclusion provides an opportunity to learn, and anyone who's successful values learning. However, learning can be a very painful process, especially when it requires a shift in perspective or when it challenges traditional thinking or behaviors. I advise leaders to accept and welcome the effort required because the payoff is incredible. You're going to have an opportunity that could impact more people's lives than you had ever imagined—an opportunity that's available only by confronting realities you may not have felt comfortable discussing in the past.

All organizations have constraints—whether it's a small number of employees, limited resources, too much bureaucracy, or a natural reluctance to change. But whether you're big or small, you can capitalize on the opportunities—to be agile, to test new things rapidly, to make material impact. And there is always the opportunity to empathize and get closer to your customers.

Building a human case

Singer/songwriter Matthew West once said, "There's no substitute for the power of a personal story." Stories amplify and bring data to life, and data opens people's eyes to opportunity and potential. John Maeda, author of several books, including *Redesigning Leadership*[1] agrees. He claims that telling a story has significantly more impact than simply explaining the reason why you're presenting the numbers. Quoting a statement he heard at a leadership conference, he's convinced that "stories trump statistics." In other words, a short narrative carries more weight than a lengthy exposition.

To build your human case for product inclusion, take the following steps:

1. **Find out what real consumers are expressing—their hopes, needs, and frustrations.** You may be able to gather this information on social media, but it's much better to meet face-to-face or gather input via online surveys or focus groups. Talk to people within or (better yet) external to your organization, such as customers whose

---

[1]John Maeda, *Redesigning Leadership* (MIT Press, 2011).

needs are not being met by your organization and its competitors. For example, when I was creating the human case to build an on-demand medical app for a UX bootcamp course I took called She Designs, I spoke to several people of varying age ranges and abilities.

Don't settle for fictional user journeys or personas composed by people who think they know the consumer. Speak face-to-face with underrepresented users in person and on a regular basis. Otherwise, you will be missing the critical information and insight to build a convincing human case.

2. **Identify trends across the input you gathered.** For example, as I interviewed people for my on-demand medical app, I discovered that older users could benefit from an option to have medical care delivered to them or to have the option to video chat with a qualified healthcare provider. Here are a few of the quotes I gathered that defined a trend:

   ◦ "Public transit in my city is not very accessible."
   ◦ "It's inconvenient to try to schedule and get to my various doctors' appointments."
   ◦ "I don't have a connection with my doctor, so I don't feel safe to tell my whole medical history."

3. **Write one or more narratives that relate each person's experience of being excluded or ignored by the current products or services being offered.** When trying to secure buy-in, different narratives will resonate more deeply with different people, so try to compose a variety of narratives. I used the quotes from Step 2 to develop a narrative around a persona I created and named "Miles":

   Meet Miles. He's 62 and lives in Boston. He lives with his husband and dog. Miles needs to access healthcare providers and his prescriptions easily. His barriers are non-accessible transportation and buildings and his busy schedule. Miles likes the option of being able to have a trusted doctor come to him. He also thinks getting prescriptions delivered to him would save a ton of time! His feedback is that there should be the same set of doctors per patient, so they can build trust and history.

   When composing your narratives, try your best to adhere to the following guidelines:

- Name the subject, even if you have to use a fictional name.
- Be explicit about the subject's demographic or intersection of demographics; for example, Latina women in the Mid-West of the U.S.
- Explain the challenge; for example, motion sensors on faucets don't pick up darker skin tones, making it hard to use automated sinks in public places. (Note: This issue has been publicized by several people who have posted their findings on YouTube.)
- Explain how this makes them feel. Are they feeling frustrated? Alienated?
- Use quotations from your interviews, if you think quotes would be helpful and if you can work them smoothly into the narrative.

If you already work with consumer researchers, meet with them to understand how they do their research and discuss possible ways to get "more proximate" to consumers and ways to bring their input together to present more cohesive and impactful narratives.

Building a business case

You can build a business case regardless of your position in the organization. I didn't have a product background when I started working with teams. I didn't have a marketing or research or engineering degree. All you need is solid data, a high-level understanding of your target demographic and your organization's stakeholders, along with a passion to make the products and services your organization offers more inclusive. Assuming you have these basic ingredients, you are well equipped to serve as the lens of inclusion through which everyone in your organization views its role in product development.

Building a business case is a three-step process:

1. Discern what real consumers are saying, including their hopes, needs, and challenges.
2. Collect data to support the business case for inclusion.
3. Organize the data into an effective presentation, using data visualization tools, if necessary, for greater impact.

We build the case for product inclusion by doing the following for each major new product:

- **Identify the market opportunity for key demographics.** For example, women comprise 50 percent of the world's population and have trillions in purchasing power.
- **Recognize what real users want.** For example, many of the women who play video games feel that the gaming industry is focused mainly on classic PC/console titles that are played in "sessions" in a fixed space and time at home. While many of them enjoy that style of play, many others prefer a more flexible approach to gaming; they prefer to game their way across various genres and devices at their own pace, to suit a variety of "moments" or moods throughout their day.[2]
- **Identify the market opportunity by examining the gap between what is currently being offered and what the key demographic needs or desires.** For example, nearly 50 percent of women play video games, yet fewer than 10 percent refer to themselves as "gamers." Even fewer actually pay for games. If the industry could create an inclusive place for female gamers, millions if not billions of dollars could be made.

The business case for inclusion

—Parisa Tabriz, Senior Director and Google's Security Princess

At Google, you'll regularly hear, "Focus on the user, and all else follows." When you have ambitions to serve users around the whole world, which we do with Chrome, you have to take into account a lot of user diversity (for example, socioeconomic, literacy, accessibility, age, skin color, ethnicity, gender, and so on) to be successful and reach your full market potential. This is especially true when you consider the world's shifting demographics and the large purchasing power of users who are currently underserved by technology. As just

---

[2]https://medium.com/googleplaydev/driving-inclusivity-and-belonging-in-gaming-77da4a338201.

one example, 15 percent of the population is estimated to have some form of vision, motor, cognitive, or hearing disability, and the global market for customers with disabilities is estimated to be $1 trillion. Those numbers are compounded when you factor in temporary or *situational disabilities* such as being sick or holding a baby or large object.

Separately, when you build for an underserved use case, you often end up with a better solution for everyone. For example, curb cuts (places where curbs are sloped down to the road) were created initially to comply with the American with Disabilities Act requirements, but now provide benefits for bicyclists, skateboarders, and parents pushing strollers, and without detriment to anyone else. Another example is OXO kitchen utensils, which were originally created because the founder's wife suffered from arthritis and was having difficulty gripping other utensils, but are now generally considered great kitchen products for everyone!

I'm excited to sponsor product inclusion because it's an opportunity to both help and learn. In just over a year, I've learned of success stories and failures that have opened my eyes, and I've been able to start sharing and scaling those learnings with various teams involved in building Chrome and other products. In terms of making forward progress, we've been trying to integrate more diverse perspectives into existing product ideation, design, and testing processes, along with other aspects of the product life cycle, and learn what works well and what doesn't. Today, product inclusion at Google is a fairly nascent community of practice, so we have a lot of room to experiment and try new approaches, and I'm focused on just raising awareness and encouraging others to think about inclusion as it relates to their work and ask questions. I hope that one day we will take a more strategic, integrated, and multi-pronged approach to building inclusively, similar to how we approach security, reliability, performance, and other critical aspects of product quality.

Building products to better serve a diverse user base is not only the right thing to do, it's good for business.

Bringing it all together

When you have both a human and business case for product inclusion, you can take a balanced approach to presenting your overall case by stressing one case or the other according to what you think will be most convincing for your audience. If you balance the business and human case in a conversation with someone who leans toward business, you not only share the data that proves this is valuable work, but you also reinforce your business case with an emotional appeal, delivering the "why" behind the numbers. Conversely, if someone buys into the people-centered case but doesn't see how it correlates with their core business goals, bringing in data on the opportunity (market size, purchasing power, etc.) will help you motivate that person to embed this insight into their core practices.

Consider an example of a pitch for product inclusion that balances the human and business case. Suppose your team is setting out to create a brand of online tutorials. In the past, it has focused almost exclusively on baby boomers, and you see a need to make future tutorials more appealing to millennials. The following pitch balances the human and business case:

> Our mission is to revolutionize the way people learn. Did you know that 25 percent of millennials are online an average of 32 percent of the time or 10 percent more than baby boomers? Millennials represent about 25 percent of the U.S. population—around 80 million people. Currently, only five percent of millennials buy our product. If we could get that to 15 percent, we would unlock millions in revenue!
>
> Millennials are online a good part of their waking hours and enjoy personalization. According to a Think with Google study, 89 percent of U.S. marketers saw an increase in revenue when they had some sort of personalization.[3]
>
> Millennials are digital natives, meaning they grew up being online, so online learning feels natural. Making people feel that you understand them (and taking time to truly understand segments of the population) will lead to increased engagement.

---

[3]https://www.thinkwithgoogle.com/advertising-channels/mobile-marketing/consumer-behavior-mobile-digital-experiences/.

Building Buy-In: Top-Down and Bottom-Up

The success of your product inclusion initiative hinges on your ability to recruit people within your organization. You need to get buy-in from top to bottom. Launch a two-pronged attack to get the buy-in you need:

- **Leadership buy-in:** You need leadership buy-in to ensure accountability, resources, visibility, vision, and support. An engaged leader will champion your cause and help reduce any resistance you may encounter moving forward.
- **Grassroots buy-in:** You need grassroots buy-in for energy, amplification, inspiration, and implementation/execution. These are the people who make or break any product inclusion initiative.

Start at the top. Having leadership on board eases the challenge of convincing others down the ladder. However, if leadership is reluctant or slow to act, you may have success starting with a grassroots effort and building momentum to convince people in positions of increasing authority.

Ideally, you want to work top-down and bottom-up to achieve the best outcome. At Google, our product inclusion team has worked with grassroots leaders to co-create an environment that enables Googlers to help create inclusive products. An example of this is the affinity groups we've created for product inclusion, including those for our Hispanic/Latinx, Black, women, Asian, LGBTQ+, and Iranian Googlers. (An *affinity group* is a collection of individuals gathered formally or informally around a shared interest or common goal. These groups build community with one another and with their external community, raising awareness of cultural events and traditions within a company.) We recruit volunteers from affinity groups to serve as "inclusion champions," sharing their unique perspectives with product teams and assisting those teams with product testing. For example, when we were creating our Pride@Product Inclusion working group, I worked with Guillermo Kalen, who stepped up to be our lead for the group to understand the best way to set up the structure, the cadence of meetings, and nuances I should understand and any key events or moments coming down the pipeline.

Recognizing the business and human imperative

—Lynette Barksdale, VP of Diversity and Inclusion, Goldman Sachs

It is no longer important for business leaders to think about product inclusion—it is imperative. The race for the next billion users is on. In order to compete in that market, businesses must be able to navigate cultural differences they never faced before. A company's future products and its ability to achieve and maintain a market advantage rests solely on its ability to connect users quickly and with integrity.

Users have many options, and the days of businesses not catering to their unique needs are over. Users want to get the best value from their dollars spent, and with the ability to move to another company or product quickly they will do just that if their needs are not met. The great news is that it is possible for companies to think about product inclusion in a way that is authentic to their business. This will require that they become more open to new ideas and new people. Businesses should take a deeper look into their decision-making processes, the people they hire, how those people agree or disagree, and determine whether they have the right perspectives at the table.

Getting buy-in from execs

Convincing executives of the benefits of product inclusion is often the most challenging task. They are busy people with a lot on their minds and may be set in their ways. If the organization is meeting its goals, they may be reluctant to entertain any thoughts about changing the way the organization functions. On the other hand, if the organization is not meeting its goals, leadership may be more receptive to product inclusion, or they may be too focused on other initiatives to entertain new ideas.

To improve your chances of convincing one or more executives to champion your cause, take the following step-by-step approach:

1. **Know your audience.** What does this leader care about? What is their goal for the year? What challenges have they seen in the past? Have any past product inclusion initiatives failed? If so, why, and

how have these failures influenced leadership's attitude toward future attempts? Understanding the answers to these questions will allow you to customize your pitch to make it as relevant and compelling as possible.

2. **Develop a clear plan.** For example, identify the key inflection points on an organization or team map where you think an inclusive lens can be added and share it in advance of any meeting. For example, if a huge marketing push is planned for this year, you have a great opportunity to introduce relevant inclusion marketing principles (see Chapter 6).

3. **Garner support from champions within the organization.** If a leader knows that there is already great work being done (at the grassroots level), they will be more likely to sign on to help, because they can see results and how those results are positively affecting their organization. Reach out to a few people who may have interest and give them a highlevel lightning talk to garner support (see Chapter 8 for more about lightning talks).

4. **Have a specific ask.** Ask for what you need; for example, a budget, an all-hands meeting, or a commitment to specific objectives and key results (OKRs). (See Chapter 6 for more about OKRs.)

5. **Tailor your case (human and business) to your audience.** Create a formal presentation with facts, figures, and consumer narratives. If you can share real user feedback, even better. (For example, if you have customer satisfaction reports or feedback gathered from social media, call out a few quotes or takeaways on a slide.) Helping people "get proximate"[4] is important; they need to connect with underserved consumers and see the negative impact that non-inclusive products have on them.

6. **Structure your presentation to keep it high level, at least at first.** Leaders have minimal time, so getting to the point quickly is essential. Here's one way to structure your pitch:
   a) Deliver an engaging and relevant headline; for example, "Did you know that there are one billion more users coming online in the next few years, mostly from India, Indonesia, Nigeria, and Brazil?"
   b) Explain what product inclusion is.

---

[4]Bryan Stevenson, *Just Mercy* (New York: Spiegel and Grau, 2014).

    c) Explain why product inclusion matters. Be sure to speak to why this work is important to you. Passion is contagious, so put some of yourself into the pitch.

    d) Make your case (human and business). Present specific opportunities or challenges that product inclusion can address; for example, an underserved market or customer feedback reflecting a golden opportunity the organization is missing out on. Be sure to include one or more brief consumer narratives.

    e) Suggest possible next steps; for example, crafting OKRs or organizing a dogfooding group (internal product user group) to gather more inclusive feedback. (See Chapter 9 for more about dogfooding.)

    f) Ask for what you need the executive to do or provide in support of your initiative.

7. **Schedule a time to meet.** Ideally, you meet with one or more executives in person, giving you the opportunity to read the room, answer questions live, and add color to why you need their support. Be sure to schedule enough time to deliver your presentation and answer any questions.

8. **Deliver your presentation.** After delivering your presentation, if time permits, answer any questions. If you cannot answer a question right away, you can simply say something like, "That's a great question; I'll get back to you with the answer." If you run out of time, schedule a future meeting to address any questions or concerns.

9. **Follow up.** After the meeting, follow up with an email recap of what you talked about, the slides, and any next steps and dates, if applicable.

Getting product and business leaders on board

—Tomas Flier, Community Advisor of the Latinx ERG, and former Product Inclusion Analytics Lead

Product and business leaders usually have a hard time understanding how biases and systemic inequities work and how this is connected to the business goals. They often feel some frustration from wanting to help but not knowing how.

However, what they do understand is business, and their mindset is perfectly aligned with product inclusion. They get it, and they want more of it because they recognize how this is directly tied to the business. Having worked in different teams, I know that the one thing leaders are focused on is the user. They know that the only way to have a successful product is to be user-centric, and product inclusion eloquently shows that not having underrepresented communities in the product development process most certainly will make a less relevant product for them. Therefore, senior leaders can easily understand why, if your workforce does not reflect your users, it's harder to build a product for unrepresented users—you end up building products for yourself or for people who look and think like your employees and not for the people you are trying to reach.

Additionally, having experience building cutting-edge products, product leaders know that different perspectives are needed for real innovation. The real out-of-the-box thinking comes from those ideas that you or your close circle wouldn't think of. The best ideas are usually those spawned by initial reactions with a product or idea—reactions that don't make a lot of sense to those familiar with the product.

Because of this, product inclusion is the most compelling business case for diversity.

Starting a grassroots transformation among employees

While securing leadership support is crucial for the success of any major initiative, implementation and execution require employee buy-in. When employees buy in to product inclusion, they make the work their own, advocate on its behalf, and sometimes even extend their efforts into other projects or activities in which they are involved. Their action comes from shared excitement and belief in the importance of what they're doing.

To get employee buy-in, you must convince them to accept your value proposition. Many people will be doing this outside of their core job, at least at first, so spend time showing these employees why you need their help, what impact they can make, and what actionable steps they can take. Here are some tangible actions you can take to start building grassroots support:

- Present your human and business case to show employees the opportunities they have to transform the lives of others while improving the organization's success.
- Encourage employees and let them co-create strategy. People buy into efforts when they can share their ideas and see them come to life.
- Hold employees accountable for the one thing they are excited about advancing, and then check in with them regularly.
- Demonstrate the unique value each employee or team brings to the product inclusion process. You really can't do product inclusion without them, so let each employee or team know specifically how they can contribute. Product managers, engineers, marketers, and user experience designers (UXers) each needs to understand the role they play and the unique contributions they can make. Product marketers for example, can add dogfooding with underrepresented users to their roadmap.

Building an inclusion-conducive workplace

—*Daisy Auger-Dominguez, Founder, Auger-Dominguez Ventures, a workplace culture consultancy and advisory firm* DaisyAuger-Dominguez .com

The starting point for organizational conversations about *diversity* (the standard nomenclature across industries) and now *inclusivity* and *belonging* has generally been through people practices. Specifically, the entry point for resourcing, strategic planning, etc., has been recruiting diverse talent. But we all know that recruiting diverse talent is simply not enough. All of your employees need to feel safe, valued, and respected, and they need to believe that they have a fair opportunity to succeed in your organizations.

When employees are afraid to speak up, engagement suffers, learning moments go unrecognized, misconduct goes unquestioned, and innovations go unrealized. In my experience, companies in which teams feel safe to fail, learn, and build/create are those where the best ideas have flourished. At Disney ABC, I saw years of advancing and preparing leaders of color while building organizational capacity for diversity, equity, and inclusion (DEI) result in the development of the network's most diverse storytelling; for example, *Black-ish* and *Fresh Off the Boat*. At Google, over time I saw business-facing leaders increasingly pause, question their assumptions, and ask for diverse sets of viewpoints/skills/experiences to make products more inclusive, far reaching, and connected to diverse audiences.

Scaling through amplification Amplify your efforts to breed more grassroots support. Through amplification efforts, people not only learn about product inclusion but also see concrete examples of the work advancing their teams' core goals, encouraging them to try it themselves. Here are some relatively easy steps to amplify the work you'd like to see happen:

1. **Create a product inclusion listserv to encourage and facilitate conversations; share updates, resources, and events; and build community.** The listserv also empowers colleagues to share and elevate work they are doing. Here are a couple tips for managing your listserv:
   ○ Send a listserv email to new groups and new people once in a while to expand its reach.
   ○ Let everyone know they can opt out at any time.
   ○ Periodically announce to the current groups and any key partners (for example, your marketing department or employee resource groups) that you are refreshing the list (minimum twice a year) to keep the community growing.
2. **Identify one or more champions across the roles or functions needed to move the work forward.** For example, you may get a senior leader, a product manager, a user researcher, and a marketing manager. Champions enable you to scale without having to do all the heavy lifting yourself. They also provide voices of authority who

are better able to convey authenticity when speaking about the work in the context of their business function. This authenticity helps to engage others within each given business function, making it easier to build community.

3. **Identify a few key principles or actions you want people to adopt or take when they buy into the vision.** For example, depending on what team has enquired about product inclusion, the easy wins would be creating a product inclusion checklist or building an OKR around inclusive dogfooding. Tangible actions make it much easier for individuals and teams to participate.

4. **Create one or more tools to remind everyone of the human and business case for product inclusion.** For example, we had one of our 20 percenters (Googlers who volunteer to do product inclusion work for 20 percent of their time), Connie Chu, build a dashboard based on a design sprint. The dashboard told the end-to-end story, starting with the current demographics, proceeding to how we are reaching our users, and ending with the practices we agreed to adopt to increase our support and extend our reach. The tool you create does not have to be in the form of a dashboard; in fact, that may be too in the weeds for certain groups (such as leaders) you are trying to get on board. However, you need to use something with impact that reminds everybody of the opportunity on the table. Including some hard stats about the underserved communities you're trying to reach always helps.

Exploring different ways to spread the word   You can spread the word via groups, in all-hands meetings, or even flyers in places you may not have thought of before. Meet people where they live both mentally and physically. Find entry points in locations workers frequent, and disseminate the value proposition succinctly. Getting eyeballs on any new initiative is important and helps to begin a groundswell, with more people hearing about the work, asking questions, and hopefully getting involved.

One of my favorite ways to spread the word is with a simple flyer posted on bathroom stalls across the company. Cleverly named "Testing on the Toilet," the flyer is a tradition at Google and is used to disperse educational information globally. In addition to educating employees, our flyers help us increase our visibility and scale our efforts. Hanging flyers with a few concrete actions is free and can get the word out to multiple parts of your organization.

Figures 4.1 and 4.2 show flyers we hung in Google offices across the globe. Each flyer delivers bite-sized snippets of actionable items for people to improve their work and their lives. We saw this as an opportunity to scale visibility for product inclusion by giving the high-level concept and actionable steps teams could take.

As you embark on a mission to change the hearts and minds of both leaders and colleagues, keep in mind that your efforts must be ongoing. People have short memories and even shorter attention spans, and they tend to revert back to ingrained thinking and behaviors. Continue your efforts to spread the word while checking in with those who have already expressed an interest or made a commitment to product inclusion. Over time, your efforts will pay handsome dividends as you create a culture of empathy resulting in increasingly more innovative product design, marketing, and sales.

The benefits of having a dedicated diversity program manager

Suezette Yasmin Robotham, a former Diversity, Equity, and Inclusion program manager for Google Search, has been integral to getting buy-in across the organization—from getting the right leaders in a room to helping evangelize and getting more 20 percenters on board. Having a person with deep knowledge of the players in the organization to help craft the narrative is integral to getting buy-in.

Suezette works to ensure that the team takes both a top-down and bottom-up approach to building inclusively. From the diversity councils she facilitates, to the diversity, equity, and inclusion speaker series she created to gain scale across the organization, she understands that the commitment doesn't just come from one group.

Testing on the Toilet Presents...
Product Inclusion: Build Products for Everyone

Debugging sucks.

Testing rocks.

by Annie Jean-Baptiste in San Francisco

At Google, we build products for billions of users across the world. But consider these scenarios where technology may unintentionally exclude users:

- A user applies a filter on a photo but their skin tone is lightened, reinforcing bias against people of color.

- A user goes to search for information about great scientists, and mostly men appear in the results.

- A user attempts to create a new product login, and there are only binary gender options.

Product inclusion (go/product-inclusion) is the practice of applying inclusive cultural & community insights throughout the entire product development process, from initial design through launch. The goal is to achieve product excellence and grow our business by building for everyone, with everyone.

Product inclusion accounts for multiple dimensions of diversity, such as:

Product inclusion applies the inclusive design paradigm to address user needs across many dimensions of diversity.

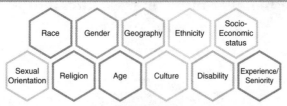

To make your product more inclusive, bring in diverse perspectives at key points in the product development process. Below are some common practices. For more best practices, visit our action planning site, or see a checklist of common issues at our site.

- Conduct inclusive dogfooding by reaching 1,500+ underrepresented dogfooders. You can become an inclusive dogfooder as well.

- Improve the quality of your UX for people with disabilities including checking contrast, providing alt labels for screen readers, and more.

- Mitigate unconscious bias in your ML training data by learning about ML fairness.

- Implement equitable UX research practices by using diverse sample data and finding internal experts on inclusive research.

Teams across Google have benefitted from practicing product inclusion, including Assistant, Apps, Docs, Arts & Culture, and more. For other success stories, visit our site.

Figure 4.1   A sample "Testing on the Toilet" flyer

Learning on the Loo Episode 206
Building for All: Integrating Inclusion

by: Annie Jean-Baptiste (Office: SFO)

At Google, we talk a lot about building for all. Whether it's through the amazing work done by our Accessibility team, or our Product Excellence work, we know that many of our users may look, think, and live very differently than we do

To ensure that our products represent the diversity of our users, the Diversity team has been working alongside several teams across Google on Project I2 (short for Integrating Inclusion). The mission of Project I2 is to influence executives and other Googlers to integrate inclusion by applying an inclusive lens to create better products and grow our business. There are some powerful incentives for integrating inclusion.

Proactive inclusion in product development can help us design useful, accessible products for large, untapped audiences.

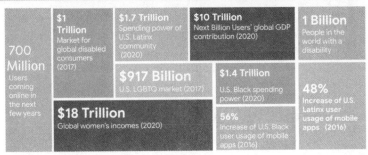

Diverse teams produce better products, and so we want to further integrate inclusion into our processes. Effectively managed diversity initiatives increase employee satisfaction. Better problem-solving and increased creativity are positively associated with a variety of diversity attributes. Research shows that mirroring the community can lead to a boost in productivity, customer satisfaction, and earnings.

If you'd like to make your team's processes more inclusive:

- Promote a diversity of backgrounds and perspectives on your core team.
- Commit to concrete actions to integrate inclusion into your workflow.
- Do a Design Think/Design Sprint and prototype ways to integrate inclusion into your business or product.
- Ensure a diverse pool of testers for dogfooding. The I2 team can help with this.

Figure 4.2   A sample "Learning on the Loo" flyer

CHAPTER 5

Adopting Product Inclusion Principles to Guide Your Work

Principles are fundamental truths or assumptions that form the foundation of a system of belief or behavior. For example, participants in every community have a set of shared principles that enable everyone to live peacefully together while pursuing their own individual happiness and fulfillment—principles such as equal justice for all, honesty is the best policy, and treat others as they would like to be treated.

Every business has certain principles in place to influence how people in the organization think about and perform their work. Examples of business principles include: the customer is always right; the bigger the risk, the bigger the reward; necessity is the mother of invention; and so forth. Sometimes, these principles are reflected in the organization's policies, practices, and mission statement.

Principles function as the North Star for organizations, teams, and individuals, providing everyone with unchanging guidance on how to think and act and insight into the reasons behind what the organization or team does and how. Principles also give work meaning, provide the means to hold people accountable, and serve as a source of inspiration.

In this chapter, I present some of the principles behind product inclusion, offer suggestions on how to develop your own product inclusion principles, and provide guidance on how to put principles into action along with an example of principles that govern Google's work in machine learning and artificial intelligence.

Checking Out Existing Product Inclusion Principles

When setting out to define product inclusion principles for your organization, group, or team, you do not need to reinvent the wheel. People working in this field have already come up with a number of product inclusion principles you can use to govern your work. Here are a few product inclusion principles you may want to consider:

- **Naming exclusion unlocks the potential of inclusion.** Just as questions are necessary to find answers and problems are necessary to find solutions, naming exclusion is necessary to identify opportunities to make products more inclusive. When looking for ways to make products more inclusive, look at how the product is currently exclusive to underrepresented groups or individuals. (Kat Holmes has done some wonderful work on this.)
- **Team diversity is reflected in product diversity.** Teams that have diverse representation or that bring in a diversity of perspectives from outside the team have the opportunity to create products that are more sensitive to the needs and preferences of consumers who are like them.
- **Everyone is different.** We all have differences in needs and preferences. As fellow human beings, we have a responsibility to make everyone feel accepted, welcome, and included.
- **Needs and preferences change with context.** The situation or context in which someone uses a product can alter what the user needs at that point in time. For example, a sensor designed to monitor the heart rate for a patient with a heart condition can also benefit people who want to monitor their heart rate for exercise.
- **Everyone is biased.** Bias provides us with mental shortcuts that enable us to function more efficiently, but bias also causes us to

overlook details, be less sensitive to other people's needs, and miss opportunities. Bringing more perspectives into the product design process helps to overcome the limitations of bias.

- **Equal is not always equitable.** Increasing access to products, so they are available to more users does not mean that more users will be able or willing to use those products. To make the products equitable, they need to be modified or redesigned to accommodate the unique needs and preferences of historically underrepresented users.

- **Designing for a minority also benefits the majority.** Products designed for underrepresented users often have additional features or functionality useful and appealing to majority users. For example, a study conducted by Ofcom showed that out of the 7.5 million people in the UK who used closed captions, only 1.5 million were deaf or had hearing loss. Closed captioning was popular among people for whom English was a second language; for watching shows in which the dialog was spoken very quickly or with accents, mumbling, or background noise; for watching videos in sound-sensitive environments, such as offices or libraries; and more.[1]

- **Diversity accelerates and amplifies learning and innovation.** If an organization is not learning, it is stagnating, and diversity accelerates and amplifies learning by bringing different perspectives into the conversations that drive innovation. People from different backgrounds are more likely to challenge the status quo, question assumptions, and present ideas that are outside the mainstream.

- **Diversity and inclusion are good for business.** This fact is no longer a secret. Business and product owners are well aware that diversity and inclusion are essential for business survival and growth in addition to being the right thing to do. Organizations that want to capitalize on the growing opportunities present in historically underrepresented populations must think and practice product inclusion across the organization.

---

[1] https://www.3playmedia.com/2015/08/28/who-uses-closed-captions-not-just-the-deaf-or-hard-of-hearing/.

Product inclusion principles at I&CO

—Rei Inamoto, Co-founder, I&CO

When we started our company three years ago, we took the time to write down a few thoughts, which then became our maxims. A few of our company maxims are relevant to the topic of product inclusion :

1. **When in doubt, subtract.** "Make it simple" is easy to say, but hard to do. So how do we make things simple? The temptation to add more creeps up when we aren't sure. Adding more will lead to more complexity. Resist that temptation. Have courage and conviction to subtract in order to relentlessly pursue the beauty of simplicity.

2. **Never say "No" without offering "Yes."** When trying difficult things, saying "No" is the path of least resistance. It's easier, smoother, and possibly more convenient. But that creates a dead end. It won't lead to new things. If we think it can't be done, find a way to say "Yes" and do it.

3. **Be tough, not rough.** Betterment requires honesty. Honesty requires toughness. There is a subtle but distinct line between tough and rough. Being tough means having a high standard, keeping ourselves to it, and being honest with each other. It means being determined to make our work better, make each other better.

4. **Risk nothing, change nothing.** Taking a risk often makes us uncomfortable. It's natural to go the other way. But comfort leads to complacency, complacency leads to stagnation, which is the exact opposite of change. The only other option to taking a risk is not taking any. And without taking one at all, there is no progress. That is precisely the biggest risk of all.

5. **Seek the invisible.** Truths and insights often lie beneath the surface. They are unclear to the mind and invisible to the eye at first, hidden between lines and numbers. Pursue the unseen in

order to uncover something that wasn't obvious before, to make something that should have already existed—for the first time.

6. **Quality is a habit.** Apparently, it takes only 21 days for something to become a habit. Put another way, it takes only 21 days for compromise to become a habit. Do we want quality or compromise to be our habit? There is only one choice.

7. **Magic > Logic.** Think about your work. How do you make this a habit? How do you use honesty to get your work to the next level? These are things we should all be thinking about.

Any organization can take these principles and put them into practice. As it relates to product inclusion, distilling down to the core challenges, building on other's ideas, seeking the invisible (making sure voices that may not always be heard are brought into the center of your work) and building a habit all resonate.

Pinpointing Your Own Product Inclusion Principles

You are certainly welcome to adopt any or all of the product inclusion principles presented in the previous section, and I encourage you to adopt at least a few of them. However, this list is not exhaustive, and you may have better ideas for principles that are more conducive to your industry, your organization, the products you build, the people who build them, and the people for whom you want to build them.

To seek out inspiration for your own ideas, take the following steps:

1. Identify areas in your organization or on your team or points in your process where you do not yet intentionally include underrepresented perspectives. Do your product teams have diverse representation? Do they bring in testers who can offer unique perspectives? What does your marketing team do to increase its awareness of underrepresented users? This exercise could lead to a principle such as, "Increasing diversity of representation across the product design and development process makes products more inclusive."

2. Identify the demographic you are trying to reach and describe what has happened historically regarding users in this demographic as related to your product, service, or industry. For example, a company that builds tools may recognize that women have historically been overlooked. As a result, it comes up with the following product inclusion principle, "Excluding 50 percent of the population is bad for business. We commit to intersectional gender diversity throughout our process."

3. Consider the moral imperative for product inclusion. For example, suppose your company creates water filtration equipment. You see that approximately 11 percent of the world's population has no access to clean drinking water, and your company would like to do something about it. This could lead to a principle such as, "Products for life's essentials should be accessible and affordable to everyone, regardless of location or socioeconomic status."

When developing and choosing the principles that will govern your product inclusion decisions and actions, be sure to brainstorm with people across your organization or team who represent the diverse populations you are trying to reach. Just as a diversity of perspectives contributes to product innovation, it can help develop the principles that guide your organization into the future.

Putting Principles into Practice

Principles are very valuable for planting the seeds of product inclusion in the hearts and minds of everyone across the organization, but committing to upholding certain principles is not enough to ensure positive change. Principles must be translated into action with the conscious intent of building products for historically underrepresented users.

Throughout this book, we explain how to put principles into action. In this section, we present one useful framework for structuring your efforts.

Developing people, processes, and products

People, processes, and practices must be developed and implemented to transform product inclusion principles into action and positive change.

One of the first frameworks I used with clients to start to expand the conversation about diversity and inclusion in product design and move them from principles to action was the "3 P's" see sidebar]:

- **People:** The users and those who invent, design, develop, test, and market products. To put product inclusion principles into practice, organizations must focus on the needs of all people while building diversity within the organization and working to make everyone more aware and sensitive to the needs and preferences of underrepresented users. The organization's culture must also allow people with diverse experiences and perspectives to thrive, or the organization will not get the full richness of these diverse perspectives. If people don't feel safe to share, they will hold back and their lack of perspective will inevitably stifle innovation. Your organization won't reap the benefits of these diverse perspectives.
- **Process:** People + process = product, so product design and development processes must operate within a framework for integrating product inclusion into the work. For example, the product design process can include design sprints that bring in different perspectives from across the organization and from outside the organization to provide insight into the ways different users may perceive and interact with the product. The infrastructure needs to also be one where diverse perspectives are valued and recognized.
- **Product:** The product is the result of people executing a process, but it is also the ultimate goal that drives the creative process. A focus on making a product as exclusive as possible is one key to achieving that goal.

This product inclusion framework was introduced to our team by Lauren Thomas Ewing—a friend, mentor, colleague, and a member of our diversity and inclusion team at the time she encouraged us to utilize the 3 P's framework (see the nearby sidebar) regularly used in the business world. This framework has become integral to how we think about product inclusion.

3 P's is a helpful framework because it is simple to understand, catchy, and all-encompassing of the work we try to get leaders and product teams to do. When trying to influence others, having a conceptual framework and a way of presenting it that sticks in people's minds

is very helpful, especially when you are talking to people unfamiliar with the concept. This framework is also effective in conveying the important point that diversity, equity, and inclusion are not merely about principles and wanting to do good, but about translating those principles into action—putting people and processes in place that are crucial for developing inclusive products.

The 3 P's approach to introducing product inclusion to your organization

—*Lauren Thomas Ewing, HR Strategy Consultant & Program Manager, Alphabet Companies*

People, Process, Product is derived from my experience doing business process redesign and organizational change management, where the goal is to create a holistic representation of how business objectives are enabled by a number of factors that need to work in concert in order for those objectives to be met. It forces you to think systemically about the challenge you're tackling or the change you're seeking to manage.

If you were to attempt to tackle product inclusion by just looking at the design of the product itself, you may end up under-examining what's going on with those who are conceiving of and constructing the product; those who are expected to use the product; the path for how the product gets from ideation to design and development to testing and launch/iteration; and the critical events, decision points, and decision-makers that influence that path. And as you consider every single one of these components, there's an opportunity to identify and eliminate blind spots and backdoors for bias to creep in and undermine the outcomes you want to see.

Lack of understanding of the user—stemming from team members lack understanding or from user research approaches that aren't robust enough—can negatively impact the choice of product to build and how it is marketed. A process that doesn't underscore the importance of inclusive perspectives and decision-making can undervalue those who can highlight potential pitfalls early and can create missed opportunities to incorporate the product features that will really resonate with end users. And finally, regardless of how beautifully engineered a product is, it can very well fail if there isn't parity or connection between who that product is built for and built by.

The 3 P's form an easy framework to remember, and they help to facilitate the process of getting people on board. Think of them like the legs to a stool—all three are needed for stability, to keep the stool from wobbling or collapsing completely. The 3 P's are inextricably linked and reinforce and strengthen one another. I've seen teams that focus a ton on people, but they need to ensure that the process (the systems in place) is viable and welcoming for everyone to share their perspectives. Similarly, I've seen people focus on building a great product but because they don't have diverse backgrounds at the table, key insights are missed that may prevent them from achieving the impact and profitability they desire.

While it may be possible to start with one P and build from there, it is certainly not optimal. I recommend developing all three and implementing them as early in the development process as possible.

Artificial intelligence leaders prioritize inclusion

—John C. Havens, author, and Executive Director, IEEE Global Initiative on Ethics of Autonomous & Intelligent Systems

Many companies are coming together to think through what fair and inclusive artificial intelligence looks like, and as machine learning and artificial intelligence get increasingly more important across industries, being proactive about having specific principles or guidelines is important. When you think about human-centered design (putting the user and their humanity at the core of all steps in the process) and Human Computer Interaction (or HCI, where the interaction of human and computers come together) you see that there is an

(continued)

(continued)

exponential increase in how we use technology. If we don't remember to bring empathy and to center humans with a particular focus on users you may not typically consider, these users have a higher likelihood of being affected by the potential bias in your technology.

As I said in my book, *Heartificial Intelligence: Embracing Our Humanity to Maximize Machines,* humancentered design, HCI, and similar disciplines help ensure machines serve humans in ways that improve our well-being and honor our values.

We're in a time when collaboration is the only way we can utilize these technologies to ensure all humanity and the planet survive and thrive. Human-centered design provides a business case for this future to happen because without it, who or what are we designing for? More specifically, better understanding end user values means building products and services that align with their cultural and personal backgrounds, providing greater market and societal relevance.

Overall, I think it's critical to think of human-centered design not just for its business value, but for its overall societal value. AI governance should certainly favor humancentered design in terms of Smarter Cities and inclusive, participant design for law in general. When beginning from the standpoint of human rights and ecological sustainability, humancentered design provides a key way to provide long-term brand relevance while also ensuring we're able to fully enjoy these amazing technologies for many years to come.

Having principles that directly call out who you are building for and how you plan to make your products and services is paramount to meeting your product inclusion objectives. Remember, as Joe Gerstandt is quoted in the introduction to this book, "If you do not intentionally, deliberately and proactively include, you will unintentionally exclude." Principles provide a means for ensuring that thinking, decisions, and actions are intentional and deliberate, and they help bring everyone involved back to the common goal of building for everyone, with everyone.

CHAPTER 6

Integrating Product Inclusion with Your Work

Implementing any new initiative in an organization can be a challenge, and product inclusion is no exception. If you have managed to convince people in your organization that it is the right thing to do and will be good for business, you are more than halfway there. Changing minds and culture can sometimes be more difficult than taking the steps required to implement inclusive product design and marketing.

Overcoming an implementation challenge is always easier when you have a framework in place—a structure to guide you along the way. In this chapter, I present three product inclusion frameworks we use at Google to facilitate product inclusion integration across all of our teams:

- **Objectives and key results (OKRs)** define the goal or destination the team has established—what it wants to accomplish in terms of product inclusion—and how and when the objectives will be met.
- **Touchpoints** identify the stages in a process that can benefit from an inclusive lens.
- **Product inclusion checklist** serves as a map to the chosen destination, detailing what the team needs to do to move from where it is now to the destination defined by the OKRs.

With these three frameworks in hand, you will be well equipped to lead your organization and individual teams through the process of integrating product inclusion into their work.

Holding Teams Accountable with Objectives and Key Results (OKRs)

Objectives and key results (OKRs) is a widely adopted framework for defining and tracking business objectives and outcomes:

- **Objectives** are memorable qualitative descriptions of what you hope to achieve.
- **Key results** are quantitative measures (metrics) for gauging progress toward achieving the stated objective.

The development of this framework is generally attributed to former Intel CEO Andy Grove, who introduced this approach to Intel and documented it in his book *High Output Management* (Vintage, 1995). Highly successful venture capitalist and author of *Measure What Matters* (Portfolio, 2018), John Doerr, who started his career at Intel and proceeded to invest in companies including Google and Amazon, introduced Google to OKRs. This framework is now used across the entire organization.

Doerr created the following formula for setting goals:

I will [**objective**] as measured by [**set of key results**].

Fill in the first blank with your objective (*what* you want to accomplish) and the second blank with your set of key results (*how* you will accomplish your objective):

OKRs serve two important purposes:

- To compel teams to define concrete objectives.
- To hold teams accountable for meeting their stated objectives on time.

Ideally, an organization has its own overarching OKRs, while individual teams have OKRs related specifically to the work they are doing. All team OKRs should support the organization's OKRs to ensure alignment with the organization's mission. At Google, our product inclusion team sets inclusion OKRs for the entire organization. We also encourage and facilitate the process of developing product inclusion OKRs with all teams across Google, especially our product and marketing teams.

OKRs are not an essential component of product inclusion, but your organization should have some framework in place for setting objectives, measuring progress, and holding teams accountable. I strongly recommend adopting OKRs.

Defining objectives

When developing OKRs, start by defining your objective—*what* you want to accomplish. Here are a few examples:

- Expand our product to Nigeria by 2020 with an 86% CSAT rating
- Improve functionality for visually impaired users
- Increase diverse representation in our advertising
- Design clothes for people of all sizes

You can make objectives as broad or as narrow as desired. Adding a date, as in the first example, more narrowly defines the objective and may help to reinforce accountability. However, you can choose to hold off on any quantitative measures because these will be established when you specify your key results (in the next section).

You can also start with a broad objective and break it down into sub-objectives. For example, "Increase diverse representation in our advertising" is very broad. You could break it down into more narrowly defined objectives, such as the following:

- Create a diversity, equity, and inclusion (DEI) advisory council
- Evaluate current advertising for DEI
- Develop ads that convey our commitment to DEI

You may want to give teams some flexibility in developing their OKRs. Developing OKRs is often more art than science. Some teams have very narrowly defined OKRs, whereas others have broader, more visionary ones. Some have annual OKRs, others have quarterly OKRs, and some do both. You may need to experiment to figure out what works for you and for each team in your organization.

Defining and scoring key results

Key results reflect *how* the objective will be achieved and how the team will know that the objective has been achieved. When defining key results, focus on quantitative measures (metrics), such as dates, quantities, and frequencies. Here are a few examples:

- Kick off robust research in Nigeria at the beginning of Q1.
- Test store prototype with 100 potential users in Lagos by end of Q2.
- Launch a Nigeriabased brick-and-mortar store by end of Q4.

Key results must be concrete and measurable. On the specified deadline, you should be able to look at a key result and answer "yes" or "no" to the question, "Did we accomplish this?"

Although key results focus on outcomes, don't wait until the deadline to look at your OKRs. Instead, use them to track your progress along the way by "scoring" your OKRs. You can score monthly, quarterly, or over a longer time frame depending on the timeline or deadline. For example, if you have an annual OKR, you may want to score it quarterly. For a quarterly OKR, consider scoring it monthly. To score your OKRs, do the following:

- Evaluate what your team is currently doing to ensure that all your team's work is aligned with your OKRs.
- Evaluate your progress toward delivering on your key results—are you a third of the way there? Halfway? Haven't even begun to think about it?
- If your team is falling short of meeting one or more key results, explore why and make adjustments, if necessary.

See Chapter 11 for details about product inclusion metrics you may want to track to ensure that your team is meeting its product inclusion objectives.

Modifying OKRs

OKRs are not carved in stone. They are tools to help monitor and manage work and ensure alignment of activities and resources to optimize outcomes. OKRs may need to be adjusted to make them more or less ambitious or to accommodate new responsibilities or tasks.

That being said, you should not be changing OKRs frequently within the time period you set out to measure. If you are doing quarterly OKRs for example, you shouldn't be changing the OKRs every week or retrofitting OKRs based on accomplishments your team has made after OKRs are set.

OKRs should be challenging, but not completely unattainable. If you're easily meeting 100 percent of your OKRs, consider setting more lofty objectives. Conversely, if you are achieving only 5 percent of your OKRs, consider expanding your time frame or breaking them down into more narrowly defined OKRs. Too lofty OKRs can chip away at morale—you want to challenge your team to dream big but also encourage and reward them with a sense of accomplishment.

If a new opportunity or responsibility arises outside the focus of your team's existing OKRs, avoid any temptation to pursue it immediately. New ideas and opportunities can be seductive and throw a team off course. Team discussion is needed to determine whether, when, and how to incorporate the new opportunity or responsibility into the team's existing workload. In other words, the team or team leader needs to prioritize the work and adjust the OKRs, if necessary.

This caution applies to product inclusion initiatives, as well. Like any new idea, product inclusion can get people very excited and eager to implement it, but consider it in the context of your team's existing work. Is it critical and feasible? Will the opportunity make the pursuit of it worthwhile? You may want to include your inclusion champions (employee volunteers from diverse communities) in these discussions. Without asking people who represent the consumers you are trying to reach, prioritizing can be difficult, and any bias can lead to wrong choices. By consulting with underrepresented users, you can more effectively gauge the importance and value of taking on any new product inclusion initiative.

Integrating Product Inclusion at Key Touchpoints

One of the first steps toward integrating product inclusion into your work involves identifying and modifying your *product inclusion touchpoints*—the stages or steps in the product design and development process where inclusive thinking and practices can make a positive impact.

Google is a technology company. As our product inclusion team worked with various product teams, we identified four key touchpoints, which align with the four phases of our "standard" product design and development process:

- Ideation
- User experience (UX) research and design
- User testing
- Marketing

Additional touchpoints exist within these four phases, but these are the four phases in the process that our product inclusion team keeps coming back to—the four points in the process where our product teams need to apply an inclusive lens. However, touchpoints vary depending on the nature of the work being done and the product or service being built.

In this section, I provide some guidance on how to identify your product inclusion touchpoints and present some examples of how to apply an inclusive lens at each touchpoint. I also stress the importance of integrating inclusion throughout the product design and development process.

Identifying your product inclusion touchpoints

To identify your product inclusion touchpoints, closely examine any and all processes in your organization that create, present, or deliver products or services to customers or the people who ultimately consume your products or services. Consider points in your processes that would benefit from customer feedback. These may be points in a process where you currently consider customer feedback but not from historically underrepresented communities.

Most technology companies could start with the framework we use at Google—ideation, UX research and design, user testing, and marketing. Businesses in the fashion industry could use a similar framework—ideation, design, and marketing—but add sample creation and editing, for example. However, some businesses have very different processes in place and hence very different product inclusion touchpoints. Here are a few examples:

- A brick-and-mortar retailer needs to consider store design and layout, location, the people and processes involved for choosing products to sell, staffing and training, product displays and advertisements, and accessibility for shoppers with disabilities.
- A doctor's office needs to consider location; hiring for diversity; language barriers; cross-cultural training for front office staff, doctors, and nurses; and diagnostic and inclusive treatment options for patients from traditionally underrepresented communities.
- Pharmaceutical companies need to look at ways to expand research and studies to underrepresented groups, increase diversity among researchers, identify underserved communities, and inform doctors concerning differences in medication response and side effects related to different populations.
- Entertainment companies need to consider hiring across multiple dimensions of diversity in every aspect of the production process, from deciding what shows to create to producing, directing, writing, casting, filming, choosing music for soundtracks, and so on.

Specifying inclusive actions at each touchpoint

After breaking down a process or business into touchpoints, specify actions to be performed at each touchpoint. Think in terms of who, what, and how you are going to make the process more inclusive at each touchpoint. Below are examples of actions for each of the product inclusion touchpoints we use on Google product teams. Actions will vary depending on differences in touchpoints and processes at each touchpoint.

Ideation Most important during the ideation phase is to gather a diverse set of participants to think through and intentionally expand the definition of your target user and the use cases for your product. For example, if you are building a product for mothers, think about who can be a primary caregiver these days—mothers, yes, but also fathers and perhaps grandparents. Some households also have two parents of the same gender, for example. The definition of target user can quickly change from mother to parents to any caregiver, and caregivers can differ in terms of race, ethnicity, gender, age, and other diversity dimensions.

By asking "who else?" you can begin to widen your target audience and bring additional perspectives into the fold to be more inclusive.

Here are a few concrete actions you can take to integrate product inclusion into the ideation phase and expand your market:

- Include people on your research team from a broad range of historically underrepresented groups.
- Talk with people in historically underrepresented communities to find out what they think about existing products, services, or brands similar to yours.
- Conduct a focus group with members of underrepresented communities to identify unique needs or preferences that need to be considered.
- Expand your research network through the use of social media, online surveys, and polling at public venues that attract diverse crowds.
- Bring in people from teams other than yours, such as marketing or human resources.
- Involve inclusion champions in the ideation process (employees, family or friends of employees, or existing customers).

UX research and design During the UX research and design phase, the purpose of your research shifts from identifying historically underrepresented consumers and their needs and preferences to determining how their differences may impact product design in terms of features and functionality. At this stage, your team will be creating mockups or prototypes and testing them on consumers in the targeted demographic.

Here are a few suggestions of how to make your UX research and design more inclusive:

- Conduct an product inclusion sprint (see Chapter 7 for details).
- Deliver a lightning talk to your product team to open minds to the idea of building with an inclusive lens (see Chapter 7 for details).
- Build a mockup or prototype and seek feedback about it from members of the underrepresented communities you want to reach.

- Obtain feedback on designs or prototypes from inclusion champions or existing customers (see Chapter 9 for details).

User testing   User testing provides an opportunity for the design team to see how their ideas play to a real, live, diverse audience. During user testing, majority and minority consumers try out the product to identify problems or oversights (what's missing) in the design. In some cases, the testers try to break the product to reveal its flaws and limitations. Even if testing does not result in a huge aha moment, diverse perspectives often lead to innovation and opportunities.

Here are a few ideas for making user testing more inclusive:

- Test the product internally with a diverse group of inclusion champions (see Chapter 9 for details).
- Conduct remote testing with a diverse group of users via an online service such as UserTesting (see Chapter 9 for details).
- To obtain feedback from more diverse groups, consider conducting testing at various malls or well-attended events.
- Create a diverse pool of product testers by encouraging people from historically underrepresented communities to register online.

Marketing   Marketing involves showing people how your product can improve some aspect of their lives. It is about telling stories that resonate with the audience—as diverse an audience as possible. Although you certainly cannot reflect the diversity of humanity in a single advertisement, commercial, or marketing campaign, you can reflect diversity across your brand or product line by presenting more inclusive stories and images.

As diversity increases among consumers—due to globalization, shifts in demographics, increasing purchase power of historically underrepresented groups, and so on—connecting with a diverse group of consumers and telling their stories becomes more crucial to the success of a product or service. Actual stories resonate due to their palpable authenticity.

As you work toward making your marketing efforts more inclusive, consider the following suggestions:

- Hire people from all different backgrounds, and make sure your hiring is reflective of the world around you.

- Have real people tell their stories about using the product. Involving consumers in the storytelling process is a cost-effective way to increase diversity, equity, and inclusion in marketing.
- Invite your inclusion champions to participate in meetings about new marketing campaigns and materials.
- Have marketing campaigns and materials reviewed internally by your inclusion champions (see Chapter 10 for details).
- Go beyond casting when creating marketing content to consider the following in the context of your target audience:
  ○ Who is directing or filming the commercial, shooting the photos, or building the website?
  ○ Who is crafting the narrative or writing the copy or script?
  ○ Who is doing the voiceover?

Applying an inclusive lens throughout the process

As far back as I can remember, I have played sports—field hockey, basketball, dance, and track and field. I think of product inclusion as a 4×100 relay race—a track event in which each person on a team of four runs 100 meters for a total of 400 meters (about a quarter mile). Finishing the race with a competitive time requires that each person on the team turns in a good performance for her leg of the race. If one runner performs poorly, it negatively impacts the team's overall performance.

The same is true with product inclusion. If you have four major touchpoints, the team responsible at each of the touchpoints must apply an inclusive lens to their work to ensure an inclusive product and customer experience.

Although every touchpoint is of equal significance, I stress the importance of starting strong because a strong start not only makes the process easier going forward but also inspires subsequent teams in the process to perform well. I encourage product teams to start strong, "right out of the blocks." (*Starting blocks* are devices used by the first runners in a relay race to push off of when signaled to start running.)

My track coach, Richard Buckner, gave me some advice that has stuck with me for life. "Ann," he'd say, "the person who gets in the blocks last controls the race." I'd look at him and roll my eyes, but I came to realize that he was right. When any of my teammates would run the first leg of the race, they would take their time doing stretches they didn't need, wave to someone they didn't know in the (small) crowd, and take their leisurely time getting in the blocks. They would then feel confident and push off the blocks with power. They were patient with the process and knew that their preparation beforehand would lead to a more successful outcome.

Likewise, with product inclusion, being intentional and diligent and taking the time to gather information and insight about underrepresented consumers is crucial to starting strong. It enables you to start with confidence and power and build momentum, which carries through subsequent stages in the process.

Admittedly, you can win a race or build an inclusive product without a strong start. You can make up time and correct for errors at subsequent touchpoints, but when you start strong, your team has a greater chance to finish strong. In the case of product inclusion, a strong start tends to result in a more inclusive product that requires less time, money, and other resources to build.

As the fourth runner of the 4×100, the "anchor," I can tell you that seeing my teammate start strong out of the blocks got me energized for a strong finish. I felt less pressure to scramble to make up for lost time. I could already envision crossing that finish line successfully for my team. Similarly, with product inclusion, a strong start in the ideation phase helps to establish a uniform flow of inclusion throughout the process, preventing one-offs, retrofits, or a scramble at the end to fix a product that is fundamentally flawed.

The earlier in the process you begin to apply an inclusive lens (for example, during the ideation process), the more value you will extract from product inclusion, thus optimizing results for both product and consumer. If you wait until the last leg of the race (the marketing phase), try as they might, marketing won't be able to make up for the lack of inclusion in the product's design, and the resulting disconnect will be glaringly apparent to consumers from historically underrepresented groups.

With a focus on all key touchpoints in the product design and development process, your teams will embed product inclusion design into how they do their work. As a result, teams will be less likely to forget or "drop the ball" (in terms of product inclusion best practices) when priorities change or the team is stressed for any reason. Everyone is now held accountable. Product inclusion becomes a part of the team's DNA.

Staying on Track with the Product Inclusion Checklist

Product inclusion is not merely a box to check. It is a mind-set and a culture of commitment to building products and services for underrepresented users. However, having a checklist of questions to ask and tasks to perform at the various stages of the design and development process can help to ensure that you and your team are not overlooking anything important.

Liane Aihara and Laura Allen, the leads for this work in their respective organizations, built upon an early version of the product inclusion checklist that is easy to understand and highly actionable. The checklist is divided into four sections that correlate with the four phases of the product design and development process:

- **Phase I: Specification and Design** includes customer journeys, product requirements, initial research, product architecture, workflows, wireframes, design exploration, and data models.
- **Phase II: Prototype and Evaluate** includes mocks, prototypes, research, content and user experience (UX) writing, visual design, motion design, design iteration, frameworks, and back-end systems and services.
- **Phase III: Build and Test** involves design quality and polish, front-end development, build testing, release management, and quality assurance (QA).
- **Phase IV: Market, Measure, and Monitor** includes marketing, analytics, key performance indicators (KPIs), monitoring, metrics, feedback, and research.

Within each of these product design and development phases are four product inclusion topics to be considered:

- **Product (what are we building?):** Strategy, plans, requirements, and objectives.

- **Representation and culture:** Representation focuses on the under-represented users for whom the product is being built, while culture focuses on language, attitudes, and beliefs that may need to be reflected in the product.
- **Access:** Availability of products and services due to factors external to the user, such as geography or income.
- **Accessibility:** The suitability of a product regardless of the user's physical, cognitive, or perceptual abilities.

This checklist (presented in the next section) is a living document, even within the confines of Google, that anyone can contribute to, customize, and share. Google teams often use the checklist as a tool for executing on their OKRs or to create their own frameworks for defining objectives, measuring progress, and holding themselves accountable. It can also serve as a baseline for questions to ask or as a tool to identify resources required.

Using the product inclusion checklist

The primary purpose of this checklist (the purpose for which it was intended) is to enable teams to apply an inclusive lens to design and build accessible products for large, untapped audiences. To use the list, find the product development phase that best matches your team's current focus and work your way through the questions and actions in each of the product inclusion topics.

Note that the questions and actions below each product inclusion topic are examples based on products we build at Google (tech products). Some questions and actions are broad enough to cover any product, whereas many others are specifically for technology products—products that contain user interfaces (UIs) and require certain infrastructure, such as high-speed Internet. Although you may be able to use the checklist as is, you probably will want to modify it as explained in the later section "Modifying the product inclusion checklist," using the checklist presented here as a template.

Phase I: Ideation, Specification, and Design For best results, consider inclusion at the very beginning of the product design process—the ideation, specification, and design phase.

Product
- What is being introduced?
- What needs are being addressed?
- Do any product policies need changed to promote inclusivity?

Representation
- Who is this product for?
- Who might have been overlooked?
- How might this product not reflect or represent different societies and cultures?
- Does more than one language need to be supported?
- Consider the diversity dimensions and intersections of dimensions: age, ability, culture, education/literacy, gender identity, geography/location, income/socioeconomic status, language, race/ethnicity, religious belief, sexual orientation, technology knowledge/skill/comfort level.
- Obtain feedback from underrepresented inclusion champions (volunteers in your organization who represent a variety of consumers).

Access
- How might this product have limited availability and access for some people?
- How might the product, infrastructure limits, or policies exclude some people?

Accessibility
- Consider the Americans with Disabilities Act (ADA) requirements and guidelines.
- Think about going beyond usability.
- Who might not be able to use this product?
- Review your design with your team's accessibility champions.

Phase II: Prototype and Evaluate As you build a prototype of your product and begin to test it internally, the questions change to address issues and concerns that arise in the execution or embodiment of the concept.

Product

- Does the prototype include a solution for all issues and concerns raised in Phase I?
- How will the product be tested/evaluated to determine whether all issues and concerns raised in Phase I have been addressed successfully?
- Share product inclusion learnings with your team, and consider sharing broadly, beyond the team.

Representation

- Does any wording used in the product or documentation specify gender pronouns? Is using particular pronouns necessary? If so, is it inclusive?
- Is more than one language supported?
- Are there any racially or ethnically inclusive images, graphics, or avatars?

Access

- How might the product, infrastructure limits, or policies exclude some people?

Accessibility

- Are multiple input modes (speech, keyboard, mouse, touch screen) supported?
- Is there sufficient contrast of text, iconography, and focus against the background?
- Are any instructions or other text easily readable and understandable?
- Is the interface accessible to users with visual impairments? Is it easy to navigate using a screen reader?
- Use a colorblind tester or simulator to diagnose any color contrast issues.

Phase III: Build and Test   Although each phase of the product design and development process is iterative, the build and test phase is perhaps more so as you fine-tune the product for quality assurance (QA).

Product
- How are issues and concerns from Phase 1 implemented in the plan to build and test the product?
- What is the anticipated product adoption and usage? (This question can help the team expand the potential opportunity if they weren't thinking about underrepresented users earlier in the process.)

Representation
- Are a diverse set of individuals (gender, race, ethnicity, age, socioeconomic status, ability, and more) included in testing?
- Are we overlooking any populations that might fit our diversity criteria?

Access
- Are people in a range of locations (including multiple countries) included?
- How will the product be tested to ensure access to users with slow internet service?
- Can the product be built in a way to make a version of it available for lower income users?

Accessibility
- Are people with assistive technology needs (for example, those who use a screen reader, magnification, or other input or output devices) included in testing?
- Test the product with a screen reader for 60 minutes if you have not used one before. This exercise is guaranteed to reveal issues that need to be addressed. (Note that other accessibility considerations need to be reviewed, but this will give you a basic understanding of why accessibility should be prioritized if you haven't been immersed in inclusive work previously.)

Phase IV: Market, Measure, and Monitor If possible, release the product to an increasing number of users in phases to allow for additional iteration. During this time, you can market, measure, monitor, and make additional modifications to the product's design to improve its success.

Product
- How closely does product usage align with predicted usage and expectations?
- What are any unexpected or unanticipated uses or issues?
- Did inclusivity policies have a positive effect?
- Were there any unintended consequences?
- Are additional policy changes necessary?

Representation
- Do the actual audience characteristics include and reflect a diverse audience (gender, race, ethnicity, age, ability, and more)?

Access
- How are language localization features being used?
- Does the geographic location and distribution of the product audience include and reflect a diverse audience?
- Are there technology access and infrastructure issues limiting product adoption and usage?

Accessibility
- Has an accessibility audit been completed, and were the requirements met?
- Has your team's accessibility champion been consulted?

Modifying the product inclusion checklist

This checklist has gone through several iterations as various teams customized it to make it more relevant to their work. For example, a product manager on the Google Chrome team, John Pallett, pointed out that for a product manager, the need wasn't to sell the team on why this work was important; it was more to give the team clear tips to expand their market and do right by the user. John modified the checklist to make it more actionable. Here is a simplified version of a portion of his checklist:

Phase 1: Design

General Scope
- What new UI does your feature introduce?
- What problem are you trying to solve with this feature?
- Does your project rely on machine learning? If so, have you diversified your training set to account for different ages, races, genders, locations, etc.?
- What steps will you take to ensure your feature doesn't succumb to biases (for example, what metrics will you measure)?

User Info
- Who is your target user?
- Who are you intentionally excluding and why?
- What typically underrepresented user groups could potentially be addressed by the new feature?

Phase 2: Prototype and Build

Language and Readability
- Does your feature support more than one language?
- Does your feature require speech? If so, have you accommodated for various accents?
- Does your feature include wording that is easy to understand (at grade level 7 or below)?

Gender, Ethnicity, and Sexual Orientation
- Does your feature (or marketing material) use wording that specifies gender pronouns? If so, have you included non-binary options (such as "they") or can you remove references to gender altogether?
- Does your feature include any racially or ethnically specific images, graphics, or avatars? If so, have you included a diverse array?
- Have you added appropriate and inclusive alt text to accurately describe the avatars or images of people to screen reader users?

Accessibility

- Is there sufficient contrast of text against its background?
- Does your feature require the ability to use a mouse? Can you navigate using only the keyboard and see clear visual focus?
- Have you made sure not to use only color or only sound to indicate key information? (For example, to indicate an error, don't just make the text red; also include error messaging text to make this more accessible for people who can't perceive color.)
- Does your feature require vision for use? Can you navigate using a screen reader?
- Do all buttons and images have proper labels for screen readers?

Payment Methods

- If collecting payment, do you support options for multiple currencies?
- If collecting payment, do you support at least three major credit card types?

Another way teams have found it helpful to customize the product inclusion checklist is to format it as a table or spreadsheet with questions in one column, space for answers in the next column, and tips/resources in a final column, as shown below.

Phase	Topic	Question	Answer	Tips/Resources
I	Representation	Did you consult users over 50?		Contact dogfooding lead to recruit inclusion champions over 50.
II	Access	Has website been tested over slow Internet or Wi-Fi connection?		Check load time in Google Analytics.
III	Accessibility	Did you test using assistive devices?		IT has a collection of assistive devices to use for testing.

If you are creating a product inclusion checklist for your organization or team, I strongly recommend that you collaborate with others to create a version that provides the greatest value to the work each team does. Depending on your organization and how work varies across teams, you are likely to end up with a number of very different product inclusion checklists.

CHAPTER 7

Getting to Know Your Underrepresented Users

Throughout this book, I encourage you to get closer with historically underrepresented users—to "get proximate,"[1] to develop empathy and understanding of their unique needs, preferences, struggles, and frustrations—especially with respect to the products and services your organization brings to the market.

The most effective way to get close with underrepresented users for the purpose of designing products and services *for* them is to build products and services *with* them. You can make the product design and development process more inclusive in several ways, including the following:

- Diversify representation across your organization and especially within your product teams.
- Recruit volunteers from underrepresented communities across your organization to participate in design sprints and testing during product development.
- Recruit customers and others outside the organization (including but not limited to family members and friends of employees) from underrepresented communities to participate in design sprints and product testing.

---

[1]Bryan Stevenson, *Just Mercy* (New York: Spiegel and Grau, 2014).

Unfortunately, organizations of all sizes, even those with a diverse workforce, are too small to reflect the diversity of the outside world. To obtain broader feedback from more diverse populations, research is required. Several research methods may be employed, including in-person interviews, online or in-person surveys, focus groups/discussion panels, and mobile research units, to name a few.

While the choice of research method is certainly important, how you conduct the research also impacts the outcome. Keep in mind that the people you design for come from different places, ethnicities, socioeconomic positions, and abilities. With those and many other varied backgrounds and abilities come different mental models and contexts of use. As researchers and builders, we are obligated to understand these contexts as much as possible in order to create truly universal products.

In this chapter, I explain how to get to know your underrepresented users through research, ensuring that your product teams have the insight and empathy to design products and services for a much broader consumer base.

The importance of listening to underrepresented users

—*Rei Inamoto, Co-founder, I&CO*

I think for design to be inclusive, it needs to start with the process. And for the process to be inclusive, the people must be inclusive.

The most important thing to do is to listen. Underrepresented communities are underrepresented because they aren't heard by the privileged. Those in power and with authority really need to listen more carefully, act respectfully, and turn their understanding into action—that's what needs to happen. It all starts with the privileged really listening.

Building an Inclusive Research Team

If you have the resources to build your own product research team, I strongly encourage you to recruit people who have a foundational understanding of the users and experiences you want to examine.

Research team members from underrepresented communities have the unique ability to:

- Share their insights as members of the communities to which they belong.
- More readily identify patterns in what study participants say and do that might otherwise be overlooked.
- Quickly identify new opportunities in collected data.
- Make the team better suited to execute on a tight timeline, by gleaning insights faster than would otherwise happen.
- Identify gaps in the team's knowledge and experience and know when to seek collaborators to fill those gaps.

Smaller organizations may not have the resources to maintain a dedicated research team. One or two people from the product team may take on that responsibility. In these situations, the researcher would be wise to consult with one or more collaborators from the underrepresented communities being studied (perhaps internal product inclusion volunteers) when designing the study and when reviewing the research.

Think of building inclusive teams as product inclusion inside product inclusion. The ultimate goal is to build an inclusive product, and you do that by building inclusion into the people and processes that design that product. The work is iterative, so the more inclusive layers you build into the work, the greater the diversity of perspectives, and the more inclusive the outcomes will be. As a mentor and former manager Karen Sumberg stated, "Repetition does not spoil the prayer!" Constantly reinforcing the need for inclusive perspectives throughout team, process, and product (keeping a steady drumbeat) reinforces and enriches the outcomes.

Modifying Research Studies for Underrepresented Users

The process of designing and conducting research studies is beyond the scope of this book. Many research methods are available, including surveying existing research, performing user testing, conducting in-person interviews, and doing surveys. Every method involves different practices and requires different research skills. Plenty of books and other resources cover everything from the basics to more advanced techniques.

Product inclusion adds a new facet to user research because participants are from populations that may respond differently to traditional research methods and questions. When designing and conducting a research study, consider the differences you are likely to encounter and prepare for the unexpected. In the following sections, I cover several areas of inclusive research that often require special attention, depending on the type of study and the characteristics of study participants.

Keep in mind that this is not a definitive list of items for conducting successful inclusive research projects. Differences in objectives, product strategies, and participants may require additional modifications to the study design and how the study is conducted.

A Six-Step Inclusive Research Framework

Making product/user research inclusive requires recruiting underrepresented consumers to participate. Matt Waddell, Google Director, advisor to Area 120 (our internal incubator for new projects) and an executive sponsor of product inclusion emphasizes the following steps for conducting inclusive research, including the need to recruit participants from underrepresented communities:

1. Explain the purpose of the study.
2. Establish your inclusion criteria.
3. Build your sample.
4. Choose a research method.
5. Conduct your research.
6. Share your research results.

In the following sections, we explain each of these steps in greater detail.

Step 1: Explain the purpose of the study

Prior to starting any study, you should be able to explain the reason for conducting the study. What do you hope to learn from the study? For example, suppose you're a cosmetics company and you want to expand your product line to appeal more to members of the Latinx and LGBTQ+ communities. In such a case, you may have several reasons to conduct research; for example:

- To find out which products are popular among consumers in these communities and why.
- To find out how members of these communities feel about different cosmetic lines and why.
- To see how members of these communities respond to your products.
- To see how members of these communities respond to your advertising.

An effective way to explain your reasoning is to engage in *landscape analysis*—an exercise focused on finding a cohesive and consistent reason for engaging in any endeavor (including research) and pursuing development of a specific product or service. To conduct landscape analysis, answer the following questions:

- What are you building?
- Who are you building it for?
- Why are you building it?
- What core challenges are you looking to solve?
- What is the opportunity? (Have users asked for a solution to a particular problem or have you noticed a gap in the market?)

After answering these questions, you should have a clear idea of what you are building, for whom, and why, and the reason you need to gather additional data to inform your product design process.

Step 2: Establish your inclusion criteria

The next step in conducting research for product inclusion is to choose your inclusion criteria. You must decide which underrepresented populations you want to reach out to with the new product or service. Inclusion criteria include the following:

- Ability
- Age
- Education
- Ethnicity
- Geographic location

- Gender
- Income
- Language
- Occupation
- Race
- Religious beliefs
- Sexual orientation

This is a short list of the many dimensions of diversity. Add to these the intersections of diversity, and the differences among individual users increase exponentially. For example, the needs and preferences of a 30-year-old Black attorney of Ethiopian descent who identifies as a woman will differ significantly from those of a 50-year-old Black business owner from Jamaica who identifies as non-binary, even though they are considered to be of the same race. (See Chapter 1 for more about intersectionality.)

If you are struggling to identify populations that may present an opportunity for your product or service, consider taking one of the following steps:

- Conduct broad research across multiple dimensions, and then conduct additional rounds of research on more narrowly defined groups.
- Consult experts within your organization to find out which underrepresented populations they think present the greatest opportunities. For example, ask someone in marketing or sales to share their insights.

Recruiting research participants from underrepresented populations

—Matt Waddell, Director and Product Inclusion Sponsor

At Google, we aim to build products with and for the world. Our users come from different places, ethnicities, and socioeconomic statuses. With those varied backgrounds come different mental models and contexts for use. As researchers and product builders, we need to understand these contexts as much as possible so we can create inclusive products, and show up in service of customers.

It's important to acknowledge that the specifics of diversity and inclusion depend on your study goals and product uses. In one situation, it might mean that you should seek participants with varied educational achievement. Another context may privilege certain job titles or industries. In other cases, it will make sense to include participants who have disabilities, or identify as underrepresented minorities, in order to make meaningful improvements to your product.

To understand the different contexts in which consumers use our products, we must conduct research with participants who represent the populations we are trying to serve. This is why it's essential to establish inclusion criteria prior to conducting research, and to consult with internal experts to gain deeper insight into where product inclusion opportunities may exist.

After building our sample, we employ a broad mix of research methods, ranging from in-person interviews to nationwide studies, to build real empathy with users. As one example: unmoderated remote studies (via services like UserTesting or Validately) are great for obtaining quick feedback from users across a variety of demographic categories, as well as focusing your research on specific customer segments.

You'll also want to consider which method best matches your research goals, including their limitations. Using the same example as above: unmoderated remote studies often require people to use certain computers and/or phones, which can significantly impact who's able to participate.

Critically, we believe that a diverse mix of voices leads to far better products, so we're excited to share our experiences broadly, and continue to invest in this work.

Step 3: Build your sample

In research, a *sample* is a small group that is representative of a larger population. Building your sample involves identifying and gathering the people you want to participate in your study. The process varies based on the source of your study participants (internal, external, online, etc.) and the nature of your study (in-person interview, national survey, etc.). For example, for an internal study, you may be drawing participants from a pool of volunteers, in which case, you may invite people to participate via email or encourage them to register on your organization's website. If you are conducting a remote study using an online tool, such as UserTesting or Validately, you will input your inclusion criteria, and the service will help to connect you with appropriate study participants.

At Google, one of our user experience (UX) research teams recruits study participants (specifically product participants) online, allowing interested parties to register to become participants. However, this same team realizes that having people register online can make the pool of participants too uniform. To increase the diversity of its research participants, the team drives around in a van to meet people where they live, as described in the nearby sidebar.

Google's user experience research vans

—*Omead Kohanteb, User Experience Researcher*

When Google conducts UX research, our sourcing campaigns typically direct people to our website, google.com/userresearch. While we are able to get a broad range of people through sign-ups at the site, there is an inherent selection bias that comes with the fact that the people who sign up there are aware of the site and have signed up to participate in UX research ahead of time.

This is where our UX research vans come in. We take our vans to locations around the San Francisco Bay Area, and on an annual Research Tour to states outside of California. On these trips, the vans allow us to conduct research with groups that we have found to be very different from those we run studies with in our lab sessions. While the groups that come to us are still self-selected in the sense that they are comfortable approaching a Google-branded van, we have found that they are less familiar with the idea of UX research, more representative of the U.S. population, and in many cases less tech savvy than the groups that come to us in the lab.

When choosing locations to take the van, we certainly consider the inclusion criteria and the intersection of different diversity dimensions, but we go beyond that to think of practical ways to reach more diverse populations. For example, we recently redesigned the exterior of the van to make it more welcoming to a wide range of identities and we park the van in locations where store merchants or staff can easily speak to us on a break. Furthermore, we make sure to welcome and encourage anyone that comes by our van to participate in studies, especially if they are from a group that is typically underrepresented in our research.

Having our product teams get out of the office in an effort to get more diverse responses and perspectives in our research is one creative way we extend our research beyond our backyard. It does not require an intricate UX lab or many researchers. It's about meeting people where they are and allowing them to tell their stories.

Step 4: Choose a research method

After building your sample, choose the research method best suited to the study's purpose. Consider the following options:

- **In-person interviews:** Talking with someone face-to-face enables you to pick up on subtleties in communication you would otherwise miss (voice tone, facial expressions, body language, etc.). It also provides the opportunity to follow up in real time; for example: "Tell me more about that" or "Why do you prefer this mock to that mock?" The drawbacks of in-person interviews are that they can be time-consuming for both researcher and participant, and they require more coordination (location, time, etc.).
- **Surveys:** The benefits of surveys are that they enable you to reach more people faster and to scale to whatever group you want to include. The drawbacks are that participants may not be giving their full attention (for example, they could be watching TV while completing the survey), you don't have the benefit of nonverbal communication, and your ability to ask follow-up questions is limited. (However, you can

ask participants whether they would mind being contacted after the survey to answer follow-up questions.)

- **Focus groups:** Focus groups enable you to get multiple perspectives without having to schedule and conduct one-on-one interviews. However, in a group of strangers, participants may not be as forthcoming as in an interview situation. You may also get some people who dominate the conversation and others who agree with the majority even when they feel differently. Be sure to allow the quieter people to speak and draw them into the conversation.

- **Remote studies:** Remote studies (usually conducted over the Internet) are great for obtaining quick feedback from a diverse pool of users. They are fast and scalable and enable you to focus your research on a specific demographic. However, they are somewhat prone to selection bias because users are often required to use specific devices to participate.

Step 5: Conduct your research

The process of conducting research is beyond the scope of this book and varies according to the chosen method—in-person interview, remote study, online survey, etc. What is important is that you have the means to collect, record, summarize, and analyze the results, so that you can extract insight to guide your future work. Here are a few suggestions to consider when conducting research specifically in the context of product inclusion:

- Obtain initial feedback as early in the process as possible, during the ideation phase, where feedback often has the greatest impact and requires the least investment to implement.
- Obtain feedback at multiple points in the process, not just at the beginning or just at the end. Teams need time to change course in response to feedback; having feedback throughout the process enables them to make easier, smaller changes in direction.
- Seek feedback from people who represent multiple dimensions of diversity.

- Be sure your researchers and focus group moderators are qualified to conduct research with the participants who meet the inclusion criteria. For example, if you plan to ask questions, have someone who understands the nuances of the group write or at least collaborate on writing the questions.
- If you are conducting a focus group or survey that includes questions, be sure the questions are easy to understand and have them checked by multiple people to identify and eliminate any bias.
- Obtain feedback on every iteration of an idea or product.
- Remain open minded when receiving feedback. Avoid any temptation to discount or ignore feedback that seemingly goes against your assumptions or product roadmap.

For any research involving participants, be sure to specify what information will be gathered and how it will be used and shared. Consult your organization's legal counsel to have them prepare any agreements that participants need to sign, such as consent forms, privacy statements, and more. Prior to the study, inform participants that they will be required to sign certain agreements before they can participate in the study; you want to avoid any surprises for the participants on the day of the study.

Step 6: Share your results

The final step in any research study is to share the results with your team and across your organization. (Remember: before collecting or sharing data or study results, you must have proper consent from participants and sign off from any partners, both internal and external.) After organizing and analyzing your data, create a research paper to present your findings. Include the following details:

- Total number of study participants
- Number of study participants in each demographic
- Questions/issues investigated
- Study design description
- Summary of results
- Conclusions/insights

After documenting your research study and results, you can share your findings with others in your organization in a number of ways, including the following:

- Email your research paper to everyone who may benefit from it internally.
- Create an infographic with a summary of the results and post it in your product team's workspace.
- Create user stories based on your findings, and share them with your team and other people in the organization.
- Use the data as substance for a *lightning talk*—a 15-minute or shorter presentation delivered to small groups (see Chapter 8 for more about lightning talks).
- Present the data and your findings at the very beginning of your team's next *design sprint*—typically a five-day session for designing, proto-typing, and testing new products (see Chapter 8 for more about design sprints).

Remember, ideally, everyone in the organization and certainly everyone on the product team should interact with a variety of users regularly. The closer to users everyone is, the more empathy and understanding is put into the products and services the organization offers. However, depending on the organization's and team's resources, having everyone involved in research is not always possible, so it is important to have a method and the means in place to share the research.

Bringing It All Together: From Research to Product

Research is not an end in itself. It eventually needs to make it into the product. During a She Designs UX course I took to bolster my under-standing of user research, I obtained a big-picture view of how research makes it into products. Our final project was to create an app with diver-sity in mind. I created an app to focus on beauty and grooming services for underrepresented consumers—specifically LGBTQ+ and people of color because these groups are not always reflected in the beauty indus-try. To create the app, I had to combine research with action, as reflected in the following steps:

1. Research the current market for beauty and grooming apps. Did similar apps already exist? If so, what did they do well? What was missing?
2. Speak to users both inside and outside my target demographic. How did they currently use beauty products, apps, and salons? Was this something they truly needed and would utilize?
3. Create a target user based on feedback I gathered from the interviews. What did users care about? With whom did they interact? How did I want them to feel after using the app?
4. Understand the how behind the why. Who was going to be servicing the potential users? How would they specifically bring a lens of inclusion in the work they did?
5. Formulate a vision for the app. Who was it serving? Why?
6. Understand the layout. How would the interface be structured? What colors would I use? Are they inclusive of those who are color blind? How would my app work with assistive technology?

These are some of the things you can begin to think about as you expand your focus to bring underrepresented users in as a priority. Taking the time to lay out your research and design strategy will save you time and enrich your product or service.

Inclusive UX from an Expert POV

—Sharae Gibbs, Founder of She Designs

Designing for people requires a deep understanding of their needs. A strong UX team is one that values and includes diverse perspectives, backgrounds, needs, and experiences. The role of a UX or product designer is not only to understand why you are designing something, but also to understand who you are designing for.

We cannot avoid bias and truly innovate without listening to the input and perspectives of a diverse set of people. There is value in diverse voices. Ultimately, the work we do has an impact on society, but diversity is also a competitive advantage.

(continued)

(continued)

During the ideation process, we need to continue to probe and continue to ask questions, such as "Does this product include the perspectives of people who identify as neurodiverse? LGBTQIA?" and "How might the work we are doing create a lasting impact in the world?" Ultimately, the products we create will become our legacy.

Think about that for a second. We all want to leave a positive legacy. Even if unintended, we don't want to leave a legacy of excluding users personally or in our product design. We all can grow and continue to improve on how we understand our end products' effects on underrepresented users and the incredible opportunity we have to impact their lives positively.

CHAPTER 8

Integrating Product Inclusion into the Ideation Process

The ideation process is the point at which you crystalize your ideas and understand what the core concept of your product or service will be—when team members start to brainstorm around how a product or service will come to life. During the ideation phase, you identify a potential solution for your target user, build on it, and solidify it before moving onto prototyping and testing.

Ideation is a key part of the new-product development process because it's when everything begins for a product or service. This is the point prior to committing resources, defining a marketing strategy, and setting solid deadlines. Focusing on inclusion in this earliest stage of product development enables you to outline more clearly key process points where an inclusive lens is needed, thereby ensuring success while reining in unnecessary and avoidable costs. As with any other piece of your strategy and process, making changes is easier and less costly the earlier in the process it is identified. Bringing an inclusive lens to bear on your process during conception enables you to plan with inclusion in mind, thus optimizing business and user outcomes while minimizing costs and maximizing opportunities, while minimizing headaches if there are any.

In this chapter, I explain how to integrate inclusion into the ideation process by conducting *design sprints* (a five-step team exercise for designing, prototyping, and testing new products) and delivering *lightning talks* (quick presentations to educate and inspire teams about the important work being done to make the company's products and services more inclusive).

Conducting Inclusive Design Sprints

One of the best ways to integrate product inclusion into the ideation process is to conduct design sprints with participants from across the organization who represent multiple different perspectives. (A *design sprint* is a time-bound, team-based, brainstorming process that enables teams to iterate on ideas quickly with minimal risk.) Inclusive design sprints are a very effective way to gather people with a diversity of perspectives to think big and inspire one another. In addition to leading to the development of more inclusive products, an inclusive design sprint provides a golden opportunity to increase participants' exposure to how different users are likely to think and feel.

The design sprints I've mentioned above are nothing new. The method was invented by Jake Knapp at Google in 2010, then further developed at GV (founded as Google Ventures) into a tool for validating ideas and assessing product-market fit. (For details, visit https://www.thesprintbook.com/.) As originally conceived, a design sprint is limited to five days, during which time the team defines success criteria, proposes competing solutions, builds a prototype, and tests it. In practice, depending on the framework, a design sprint can be conducted over the course of any time period ranging from three to four days to two weeks.

These sprints are iterative, meaning they repeat, with each subsequent sprint building on the work of the previous sprint. Near the beginning of each sprint, the sprint team defines a challenge, and at the end it has a product or prototype that overcomes the challenge or brings the team closer to overcoming the challenge. In addition, and perhaps more importantly, each sprint results in learning and leads to additional questions or challenges to guide future sprints.

In the following sections, I provide guidance on how to prepare for and conduct a design sprint.

Identify and invite participants

To ensure a successful design sprint, make sure you have the right people in the room. You want to populate the room with a broad representation of perspectives and expertise to fill the following roles:

- **Sprint master:** The sprint master (facilitator) keeps the sprint sessions on track, ensures that all participants have a voice in the process, and records the learning that takes place.
- **User representative(s):** User reps are people from the underrepresented groups for which the product will be designed and, perhaps, others who have deep insight into the needs and preferences of underrepresented users. Ideally, you want to keep the design sprint "in the family" by inviting internal users and existing customers who represent diverse demographics or the specific groups you want to reach. (Because the ideas are new, you may want to keep them as confidential as possible until they are more advanced and thought out.) If there's an opportunity to invite external trusted testers, that's also wonderful.
- **Customer service representative or manager:** Customer service personnel interact with current customers and have unique insight into their needs, preferences, and frustrations.
- **Marketing specialist:** Marketing personnel can offer valuable input on demographics and shed light on who is being excluded from marketing efforts.
- **Product designer/developer:** The product designer/developer translates user needs and preferences into features and functionalities—a skill that is essential for exploring possible solutions to the challenges discussed.
- **Technology specialist:** A technology specialist is uniquely qualified to offer insight into the capabilities within the organization for building any product the team may envision.
- **Decision-maker:** Every sprint team should include a senior executive or a stakeholder who has the power to decide whether a product

idea will be pursued and to make the final call when consensus among team members cannot be reached.

To make your design sprint more inclusive, be sure to bring in people who are fairly well removed from the product team's usual work. You want to gather people with diverse perspectives who see the product with fresh eyes—people who have no idea about the team's work and might ask questions that uncover what otherwise would probably be overlooked; for example, an unforeseen user-experience glitch or a marketing disconnect. Try to limit the number of people to a maximum of 10 to enable all participants to have a voice. Be sure to seek out different perspectives, and notice who speaks up and who is quieter.

During the sprint, if you have instances when small teams need to be formed to complete certain tasks, form those teams with people from different departments or functions. Forming teams of individuals who do not normally work together creates innovative synergies and enables participants to build off one another's ideas without the pressure of having to adhere to department norms or strategies. It allows people to be freer with their ideas, thus facilitating innovation.

Lay the foundation for a successful design sprint

Following are a few suggestions to optimize the outcome of your design sprint:

- A couple weeks prior to the first sprint session, send an email invitation to the people on your list that includes the following details:
  - Background information on the design sprint, such as the purpose of the sprint and who will be participating
  - The date(s) and time(s) of the sprint meetings
  - A simple sign-up form, or a link to your sign-up form
- Reserve a room in advance large enough to accommodate everyone comfortably with additional room to move around.
- Prepare what you need to record and share ideas (whiteboard, sticky notes, markers, etc.).
- Encourage participants to use electronic devices solely for the purpose of furthering the mission of the design sprint (or you can choose to prohibit electronic devices).

Conduct the design sprint

The original design sprint developed by Jake Knapp is a five-day, five-step process:

1. Map.
2. Sketch.
3. Decide.
4. Prototype.
5. Test.

For an inclusive design sprint, I suggest adding two key steps, one at the beginning and one at the end to create a seven-step process. You can still run the sprint over the course of five days, but these additional steps are key to introducing the importance of product inclusion and ensuring the work continues beyond the sprint:

1. Introduce product inclusion.
2. Map.
3. Sketch.
4. Decide.
5. Prototype.
6. Test.
7. Capture learning and plan next steps.

In the following sections, I describe each of these steps in greater depth.

Step 1: Introduce product inclusion At the very beginning of the first sprint session, introduce product inclusion in the context of the sprint's overall purpose or challenge. A great way to introduce product inclusion is to deliver a brief presentation we like to refer to as a "lightning talk" (see the later section "Delivering Lightning Talks" for details). Whether you deliver a lightning talk or find some other way to introduce product inclusion, be sure your introduction includes the following:

- The definition of product inclusion—building products for everyone, with everyone

- Both the human and the business case for product inclusion (see Chapter 4)
- A statement of the importance of building intentionally for minority users—because otherwise they are likely to be excluded
- An introduction to user reps who may not be familiar to the other members of the team and an explanation of the valuable role they will be playing. This gives you a golden opportunity to make underrepresented participants feel welcome, appreciated, and free to express their thoughts and insights.

Step 2: Map   After introducing product inclusion, create a map to highlight user pain points or problems they encounter when using the product, along with opportunities that those paint points reveal. The following two mapping methods are commonly used in design sprints:

- **Customer journey map:** A customer journey is a chronological map that traces the customer's end-to-end experience with the product or service (from pre-purchase to post-purchase), highlighting problems or pain points along the way.
- **Empathy map:** An empathy map is divided into four quadrants to evaluate the user experience based on what the user says, does, thinks, and feels. The map can reflect a single user's experience or the collective experiences of multiple users. The idea is to develop a persona which drives the design process.

Next, based on insights gleaned from the map(s), brainstorm possible ways to address the problems or pain points. What is broken that needs to be fixed? What features could be changed or added to the product to improve the user experience? What functionality can be changed or added? At this point, the team should be moving from problem(s) to opportunity(ies).

Finally, identify one problem/opportunity the team deems most important to pursue.

Step 3: Sketch   Sketching involves drawing one or more solutions to the problem or challenge. Jake Knapp, who invented the design sprint, recommends the following four-step sketch process:

1. **Gather key info.** Spend 20 minutes discussing similar products already in the market and other sources of inspiration, and jot down what you find.
2. **Doodle rough solutions.** Spend another 20 minutes sketching rough ideas.
3. **Try rapid variations.** Choose your best idea from Step 2 and sketch out eight variations of it in eight minutes.
4. **Figure out the details.** Spend 30 minutes figuring out and jotting down your final solution for overcoming the problem/challenge.

Step 4: Decide   Now that you have multiple ideas for solving the problem/challenge, you need to trim the list to reveal only the best ideas. Knapp recommends the following five steps for reaching consensus on which solution(s) are strongest:

1. **Art museum.** Tape the solution sketches to the wall in one long row.
2. **Heat map.** Have each person review the sketches silently and put one to three small dot stickers beside every part he or she likes.
3. **Speed critique.** Three minutes per sketch. As a group, discuss the highlights of each solution. Capture standout ideas and important objections. At the end, ask the sketcher if the group missed anything.
4. **Straw poll.** Each person silently chooses a favorite idea. All at once, each person places one large dot sticker to register his or her (non-binding) vote.
5. **Supervote.** Give the decision-maker three large dot stickers and write her initials on the sticker. Explain that you'll prototype and test the solutions she chooses.

Step 5: Prototype   After choosing a solution to pursue, your team builds a *prototype*—a preliminary model of a finished product—which can be tested on users. Making the prototype as realistic as possible ensures the best possible results when customers test it—the next step in the sprint.

Step 6: Test   On the final day of the sprint, show your team's prototype to five customers, one at a time, ideally talking about the product and observing how they use it or interact with it. Video-record the session with each customer, so the team can watch the videos together and learn from them. Knapp recommends the following "five-act interview" with each customer-tester:

1. **Friendly welcome.** Welcome the customer and put them at ease. Explain that you're looking for candid feedback.
2. **Context questions.** Start with easy small talk, then transition to questions about the topic you're trying to learn about.
3. **Introduce the prototype.** Remind the customer that some things might not work, and that you're not testing them. Ask the customer to think aloud.
4. **Tasks and nudges.** Watch the customer figure out the prototype on the customer's own. Start with a simple nudge. Ask follow-up questions to help the customer think aloud.
5. **Debrief.** Ask questions that prompt the customer to summarize. Then thank the customer, give them a gift card, and show the customer out.

Step 7: Capture learning and plan next steps   A sprint is not intended to be a "one-and-done" activity. Sprints are iterative activities conducted for the purpose of continuous improvement. Even if a prototype fails, the sprint is always a success because it enables teams to capture learning and develop well-informed plans for moving forward. In many cases, the plan to move forward consists of scheduling another sprint. Sprints can bring up new ideas or other paths the team wants to pursue. Likely, you won't create a new product that is ready to be shipped since it is merely a prototype, and so creating a concrete plan with next steps is critical.

Although the sprint is officially over after testing, I strongly encourage you to take the following steps before wrapping it up:

1. Engage in a team discussion of what was learned during the sprint, and be sure to take notes to capture and record that learning.
2. Commit to specific action items, owners, and deadlines to ensure that your work progresses beyond the sprint.

Delivering Lightning Talks

Lightning talks are short presentations (15 minutes or less) that introduce a concept to new audiences. Many times, they come at the beginning of a sprint to open the sprint attendees' eyes to other teams, concepts, and practices that may spur ideas they can advance later in the sprint. Also, if you don't have the time or bandwidth to do a full design sprint, doing lightning talks for teams to inspire them in team meetings, brainstorming sessions, all-hands meetings, or other group settings can help people start to see their work through a more inclusive lens.

Having a lightning talk can whet people's appetites to learn about product inclusion and leave them wanting to learn more. We did this both internally and externally to test what resonated with our audiences and what did not. We received a lot of positive feedback, with both internal and external product teams wanting longer consultations, as well as consultations to curate content and ensure they were building with an inclusive lens and doing right by their customers, thereby unlocking more value. Getting product inclusion "in the water" is key to having a shared language and getting people to think more broadly about adding potential customers and shifting the conversation from one around diversity to one of increased opportunities, incremental customers, and creating more user value. Because each talk takes only 15 minutes, it is scalable, easily digestible, and can spark excitement with people who are unfamiliar with the concept.

We once delivered a lightning talk to a product team that did not yet have a product inclusion working group. We were able to snag 15 minutes of their team meeting and talk about what product inclusion was, the benefits of using our product inclusion checklist (see Chapter 6), potential outcomes of incorporating product inclusion, and next steps to get started. The team immediately created a product inclusion working group of 20 percent volunteers who would work with our team to create a strategy for their whole organization. This was effective because we had clearly laid out what the concept of product inclusion was, identified opportunities for their specific product and how it would help the user, and outlined three next steps they could take. We also shared how we would support them along the way. The team is now leveraging our inclusion champions for nearly their entire product suite. (*Inclusion champions* are employees from various historically underrepresented communities, such as women, people of color, people over 50 years old, and so forth, who share their perspectives and test products.)

The key to a successful lightning talk is preparation. You have only 15 minutes (or less) to make your case, so create a tightly organized presentation. Consider structuring your talk using the following framework:

1. Present a captivating headline or case study to draw in the participants, get them excited, and start to make the human case for product inclusion.
2. Introduce the concept of product inclusion. For an unfamiliar audience, you may want to start by defining product inclusion as the practice of applying an inclusive lens throughout the product design process.
3. Explain why it's important from a business perspective. For example, present some demographics that illustrate the business opportunity.
4. Present data and/or examples to support the importance of integrating inclusion into the product development process; for example, case studies or actual user feedback. (See Chapter 4 for more about building a human and business case for product inclusion.)
5. Suggest three to five concrete actions attendees can take to begin to integrate inclusion into their work; for example, join our listserv, sign up for our dogfooding pool, commit to following inclusive marketing principles. (*Dogfooding* involves the testing of a newly developed product by internal users; see Chapter 9 for details.)
6. Share any product inclusion resources you have (such as an internal website) and ask audience participants to commit to one action to begin to integrate inclusive design into their work.
7. Provide contact information to enable attendees to get in touch with you or someone from your team for additional information or guidance.

Lightning talks are a quick way to build enthusiasm and distill down the few key actions you'd like teams to take to get started. In 15 minutes, you can get buy-in to have product inclusion become a part of the product design process, and if you do this at the ideation phase, the benefits will trickle down through the rest of the product design process.

Remember, ideation is when the biggest, boldest, brightest ideas can come out. It's fun to think through innovative ways that can change users' lives for the better and harness the creativity and power of people in your organization to turn dreams into actions and tangible products. Whether it's an hour, a day, or a week, taking time to ideate and ideate inclusively is key to building inclusive teams, products, and organizations.

CHAPTER 9

Starting Your Own Dogfooding and Adversarial Testing Programs

C ats don't like dogfood. How do I know? The only way I could know for certain—by offering a variety of dog foods on a variety of cats (just kidding! I would never do that). But the same is true of human subjects—you cannot assume you know what people need without asking them and having them try products. Just as a dog's preferences differ from those of a cat, what works for one user may not work for another. Individuals are all different and need to be treated as such. The only way to find out what appeals to and works well for an individual or a community of users is to ask, listen, and test . . . repeatedly throughout the product design and development process.

Talking to users and having them try products are crucial for creating relevant and inclusive products. Unless you interact with real prospective users and have them interact with your products, you risk creating something that some people cannot or will not use. Being able to build for everyone means that you must actively seek out perspectives of historically underrepresented users across dimensions of age, race, ethnicity, ability, and more.

We practice various research methods at Google to develop a better understanding of historically underrepresented users. Our team prioritizes the following two research methods:

- **Dogfooding:** At Google, we believe that we need to "eat our own dogfood" before releasing it to users. Dogfooding involves internal discussion about or testing of products prior to launch. Dogfooding may involve focus groups or user testing.

   Our trusted tester program extends dogfooding to family and friends of Googlers and trusted partners that provide concrete and honest feedback and share ideas.
- **Adversarial testing:** With adversarial testing, we use a diverse group of internal testers (dogfooders) and assign them the task of trying to "break" the product before it launches. The goal is to reveal any defects, including anything that may be or appears to be exclusive of underrepresented users, so issues can be addressed prior to product launch.

Note that we refer to dogfooders and testers from underrepresented groups as our "inclusion champions."

In this chapter, I explain how to start your own dogfooding and adversarial testing programs and coordinate with product teams to engage participants in these programs in their work. I also present four case studies to demonstrate how inclusive practices in the product design and development process improve the quality of products.

Creating and Managing a Dogfooders/Testers Pool

If your organization is large and diverse, I encourage you to create a pool of dogfooders and internal testers that product teams can draw from when they need feedback from underrepresented users. At Google, our employee resource groups (ERGs) serve as a fantastic source of partnership to find volunteers who come from historically underrepresented backgrounds. ERGs are employee-initiated groups, financially and programatically supported by Google, that represent social, cultural, or minority communities. We have approximately 15 such groups, including the following:

• HOLA: Hispanic/Latinx Googlers
• BGN: Black Googler Network
• Greyglers: Older Googlers
• IBN: Inter Belief Network

We periodically invite employees across the company to join the general dogfooding pool, as well, and they get regular communication about dogfooding opportunities they can opt into, such as the following:

• A team is close to product launch and needs feedback on experience with the product.
• A team has a prototype for a product and wants to bring in diverse perspectives for testing and feedback.
• A team has identified a potential opportunity, but wants to make sure that the opportunity resonates with a broad set of potential users.

In the following sections, I provide guidance on how to create your own pool of dogfooders and internal testers and engage them in the work of your product teams.

Recruiting employees

Employees are often eager to serve as dogfooders and adversarial testers, so you should have no trouble recruiting people. However, your organization must be willing to provide employees flexibility in their schedules to permit them to serve in these roles and sensitivity knowing this is not their core role. Recruiting itself is often just a matter of asking people to serve:

• If your organization has ERGs (or something similar), collaborate with the leaders of the ERGs to recruit members from their groups. Consider delivering a lightning talk (see Chapter 8) to the group on the topic of diversity and product inclusion and have a sign-up sheet to gather the names of interested members.

- Email the entire organization explaining what product inclusion is and your aim of creating a diverse group of testers, and invite them to join your dogfooding pool.

Recruiting employees to join your dogfooding pool is an ongoing effort, not a one-time exercise. Here are a few tips to optimize your success:

- Reach out to those who typically do not have the opportunity to participate.
- Make it optional.
- Always thank people for their time, effort, and expertise.
- Share the "why" behind the product and testing, so everyone is aware of how essential their participation is.
- Be open to feedback and to implementing it; otherwise, dogfooders will begin to feel as though their input is not valued.

If your organization's workforce is too small or lacks the diversity needed to create a pool of internal dogfooders and testers, consider other options, such as the following:

- Conduct remote studies using services such as UserTesting and Validately (see Chapter 7 for details).
- Extend your recruitment efforts to outside sources, such as existing customers, universities, industry and professional groups, and non-profit organizations.
- Hire an outside firm to conduct focus groups and user testing for your organization.

Choosing a dogfooding lead

As you build your pool of dogfooders, I recommend that you choose a dogfooding lead (from the pool or elsewhere in the organization)—someone willing and able to coordinate between dogfooders and product teams. Note, you'll have to spend some time upskilling this person so they understand the remit of their role.

At Google, we have had a few dedicated dogfooding leads. Austin Johnson, who initially chose this position as his 20 percent project served as our dogfooding lead until the end of 2019 when Ben Margolin and Robin Park took the helm. He now works at YouTube as the user testing strategy point of contact and consultant for teams across Google. The point of a dogfooding lead is to work with teams to assess their testing needs, think of ways to bolster the representation in their testing across multiple dimensions of diversity, help track their progress, and hold teams accountable for integrating inclusive testing in their work.

Managing our Inclusion champions

—*Austin Johnson, Partner Operations Manager, Music Labels, Former 20 Percent Dogfooding Lead*

My role is to help scale our product inclusion initiatives by managing our list of volunteers from underrepresented backgrounds (called Inclusion Champions) and sending them opportunities from our product teams to test products and provide feedback before we launch these products externally. A majority of the requests have been from the product teams themselves asking for input from our product inclusion team, or asking for us to reach out to our volunteers to test a new app or hardware that will go to market soon.

Bringing in diverse perspectives to evaluate our products before launch can ensure that we create solutions for all of our users and that our products aren't creating a divide between communities, but actually empowering various communities of users. If we are building with only a few people and perspectives in mind, then we are not actively trying to be a part of the solution to improve the world with our technologies.

Teams benefit greatly from inclusive user testing in numerous ways. Not only does it allow people on teams to grow in ways that are more difficult to measure (such as building empathy and iterating on the creative process), it also helps people troubleshoot their

(continued)

(continued)

products and create better solutions by providing access to internal Google volunteers from underrepresented and underserved communities that may not be represented on their teams.

This can benefit product teams by ensuring that particular features are stress-tested before they're released; we certainly don't want to release a product before it's ready. For example, I've helped source volunteers to test products that needed to render all skin tones correctly. The only way to test something like this is through the use of a large sample size of people with varying skin tones. Without our volunteers, this would be much more difficult for teams to test internally.

Coordinating dogfooders with product teams

Let your product teams know who your dogfooding lead is and how to get in touch with that person. Instruct them to send a written request when they need to engage the services of dogfooders for focus groups or testing. This request must include the following details:

- A brief description of the product or service the team is creating or wants to explore.
- A brief description of any diversity and inclusion research the team is currently doing or has done in the past regarding this product or service. This is important in helping the dogfooding lead understand who the product team considers to be their core user, so the right people can be lined up for the focus group or testing. The dogfooding lead may be able to find gaps in the product team's current dogfooding pool and help them expand.
- A time frame or timeline that indicates when feedback/testing is needed or must be completed.
- Any other parameters or restrictions.
- The product team's work with other relevant cross-functional teams, such as legal and marketing (and any necessary sign off from those teams).

- A draft of a note to be distributed to your dogfooders that includes the following:
  - A brief description of the product team's purpose for engaging the dogfooders
  - The timeframe or timeline to facilitate scheduling
  - A list of FAQs to answer any questions the dogfooders are likely to ask about the study
  - Contact information for someone on the product team who can address any additional questions or concerns.

Upon receiving a request from a product team, the dogfood lead must send an email notification to all dogfooders who fulfill the relevant inclusion criteria. (Depending on the product and the feedback needed, you may want to extend your invitation beyond registered dogfooders.) When composing the email notification, specify as many of the following details as possible:

- The purpose of the focus group or test and the intended impact of the product or service
- Time frame or timeline when you will need them
- Location
- Any restrictions to ensure that only those eligible reply back
- The deadline to respond/sign up
- A statement reinforcing that their participation is voluntary

Here is a template you can use to announce your dogfooding opportunity:

Hi inclusion champions!

We have the opportunity for you to help dogfood [product name]. We're looking for colleagues in [specified location] to try this product out for the next month.

We'll need you to use this product outdoors, and we'll be sending out weekly feedback surveys on functionality, design, and your overall experience. [product background and study purpose]

Please sign up [here] if interested and we'll get back to you!

For more information, consult the study's [FAQ] or reach out to [contact].

All of us appreciate your dedication to building products for everyone!

At regular intervals leading up to the date of the focus group or the date on which testing starts, send out reminders letting your dogfooders know of the approaching date, and include the location and any other information they need, such as items they may need to bring.

Following up with the product team

Check in periodically with the product team leader to track the team's progress, gather any learnings they can share, and put together a plan for any follow-up dogfooding sessions (if applicable). Here's a sample letter (composed by Austin Johnson) that you can customize:

Hi [insert name],

I hope all has been going well!

I'm reaching out on behalf of the product inclusion team to see how things have been going since we helped you think through building with an inclusive lens. What is statistically significant? Any particular insights you've heard from dogfooders?

A reminder that these are some of the metrics that we are trying to assess by partnering with your team and other product teams:

[metrics you are tracking, see Chapter 11]

Let us know if you have any questions. We look forward to hearing back from you.

Cheers,

[your name]

Our product inclusion team sends a note at least once a quarter to product teams to ask how things are going. These notes can be short and sweet asking if they need help or have anything interesting to share, or you can share the metrics that matter to your team and ask them to share the metrics that are important to them and any shifts they have observed in the metrics indicating positive or negative movement related to product inclusion.

Email responses from team leaders keep us updated on what is going on with the teams we partner with and reveal areas where we may be able to provide additional support. This feedback also enables us to measure and benchmark progress across teams and products and ensure accountability for product inclusion (including our own).

Following up with your dogfooders/testers

In addition to following up with product team leaders, keep in touch with your dogfooders. Especially important is to follow up with dogfooders who recently participated in a focus group or product testing session. Email them to express your appreciation and let them know how the time, effort, and expertise they contributed positively impacted the product team, their work, and the product they were/are working on.

Case Studies on Inclusive Dogfooding/Adversarial Testing

Across Google, product teams reach out to our team for help with focus groups and testing throughout the product creation process—from ideation and product development to the near-launch and post-launch phases and marketing. In this section, we present four case studies of product teams that have leveraged our inclusion champions to their advantage: Owlchemy, Grasshopper, Duo, and Project DIVA.

The intentional focus of these innovative teams has helped to make products that serve the needs of underrepresented and majority users alike. As you read through these case studies, look for instances in which these teams:

- Thought of inclusion at the onset of their work
- Looked across multiple dimensions of diversity
- Kept iterating based on feedback

Case Study: Owlchemy

Owlchemy is a video game developer acquired by Google in 2017. Currently, the Owlchemy team is in charge of developing virtual reality (VR) products and avatars. By speaking to underrepresented Googlers from the moment they initiated their work, they have been able to build more inclusive products that accommodate wide-ranging differences among users.

In the following sections, the Owlchemy team discusses the development of its Vacation Simulator, sharing details about a system that enables players to bring themselves into the virtual world—Avatar Customizer.

VR for everyone At Owlchemy Labs, "VR for everyone" is at the heart of everything we do. We believe everyone should be able to play our games, have fun, and truly feel like themselves.

To live up to this philosophy, we set out to build an avatar system that allows everyone to be represented, lets players spend as much or as little time as they want customizing, and feels both believable and relatable.

Tons of research, consultation, and constant testing, testing, testing led us to the refined yet powerful set of options in the Avatar Customizer. We're thrilled to share the results of our (ongoing) work, as well as a little bit of the process behind creating a system to represent everyone in VR.

What's in an avatar customizer? When we started development, we realized that vacations are filled with activities that require you to see yourself—dressing up for your destination, checking out new outfits in the mirror, and, of course, taking selfies! For players to truly bring themselves into VR, we knew it was essential to build an avatar system beyond a floating headset and hands.

While an avatar system is a massive undertaking for any project, the nuances of developing for VR added a few layers of complexity to the task. Technical hurdles aside, the physical presence and agency of VR make it all the more important for players to identify with their avatars. True to Owlchemy fashion, we accepted this important task gleefully!

First, to account for the vast range of visual elements that make up a person's physical identity, we narrowed things down to what we found essential. We want our players to be able to capture the core elements of their physical appearance in order to seamlessly embody their virtual selves in their adventures across Vacation Island.

At the start of the game, you're greeted by a friendly and accessible interface for customization—a vanity set full of switches, dials, and knobs for tweaking your avatar's appearance. Adjust the height of the vanity with the handle, and you're off creating your virtual self!

Our Avatar Customizer allows players to customize the following:

- **Skin tone:** Skin tone is the most immediate personal identifier and one that nearly everyone considers core to their identity.
- **Visor color:** Instead of eye color, we let players customize the color of their visor. (You are in a VR simulation, after all!) The visor color can represent your eye color or be an aesthetic choice if you want it to represent your style.
- **Hair style and color:** We wanted to ensure everyone could find a hair (and beard!) option that reflected their identity—be that color, texture, style, or even religious/cultural identifiers like hijabs and dastars (an important, core part of identity for many!).
- **Glasses:** Much like with hairstyles, we know that for some people glasses are more than an accessory—they're a part of who you are!

Using just a few core features, we polished our Avatar Customizer to be a powerful tool to create a truly believable avatar in VR. We found simplicity overall to be the key to creating an iconic representation of one's self, with optional depth in only the most essential areas.

Whether you're taking a selfie on the beach...

...or trying to pick your favorite accessory in the Dock Shop...

We want you to feel like...well, you!

The process and considerations For "everyone to feel like themselves" in our game, we really meant *everyone*. We feel a great sense of responsibility to pave the way for a more diverse, inclusive baseline for implementing avatar systems in VR. Avatar systems in games have often failed specific minority groups in the past, and alienating someone in VR with a half-baked system carries even greater consequences than in traditional games.

Everyone deserves to see and be themselves in the world—including virtual worlds. Because of this, we committed ourselves to the entire scope of an avatar system. We knew that we had a lot of work ahead of us, but not how much until we started to dig deeper.

Our first step was gathering as much information as we could on our own. We took skin tone inspiration from artist Angélica Dass's Humanae Project[1] and Fenty's foundation line.[2]

With hair, we started at the texture level, learning about the 1–4C classifications for hair textures from a variety of hair and beauty websites (1: straight, 2: wavy, 3: curly, and 4: coily or kinky). Then came styling. For this, we looked at the most common haircuts across age, gender, texture, and major cultural groups and tried to represent two to four styles from each group.

Just this research gave us plenty to start working on, but we knew that we needed additional expertise to provide context to our existing knowledge and fill in the inevitable gaps. We reached out to fantastic groups including Google's product inclusion team, our local experts from Pretty Brown & Nerdy, and even a few of our industry friends (thanks, Rami!)—all of whom helped us immensely to spot major gaps in our representation and understanding. Using external help, we addressed things like a lack of hairstyles for older women and long-haired men, appropriate terms for Muslim headwear, and inaccuracies in our textures and models for curly and coily hairstyles.

Sometimes, representation even meant developing new tech. When tackling Black male hairstyles, we implemented a method for transparency gradients in hair to capture the look of the fade. We also wanted longer hairstyles to move with the player, so we developed custom physics tech for long hair to feel more realistic. For the hijabs, we had to create a new animation rig so that the hijab would stay anchored to the torso and never reveal the skin of the neck.

These details, while seemingly small, were important for us to implement in order to truly allow everyone to represent themselves. Overall, we estimate that the Avatar Customizer in Vacation Simulator has taken us over 2,000 development hours to date, and we're still counting!

---

[1] https://www.angelicadass.com/humanae-project.
[2] https://www.fentybeauty.com/face/foundation.

Be Who You Are in VR! We're delighted to show the richness of our avatar system and its customization options, as well as the process behind the scenes that led us to the system we have today. It was no small feat to tackle a challenge of this scale with a VR project, but it is one we felt strongly about and committed to getting it right. After all, *everyone deserves to be themselves in VR!*

We can't wait to see all the ways players customize their avatars in Vacation Simulator when the game launches! Kick back, relax, and get ready to book your vacation!

Commentary What I love about this case study is that Owlchemy prioritized product inclusion from the very start and built inclusively from the onset, instead of having to find inequities later in the process and patch them. By securing a group of inclusion champions across multiple dimensions of diversity—including gender, race, and religion—the team was able to ask questions, obtain feedback, and iterate in real time to make the product more inclusive. They learned about authentically representing people and sought feedback on everything from skin tone and hair texture and more. This, in my opinion, is the ideal way for teams to function.

Nearly all product teams engage in some sort of user research and testing. Bringing diverse perspectives into that research does not require a huge additional investment, but it often delivers insights that product teams may never have otherwise imagined. Inclusive research sparks creativity and innovation and can clear the path to previously inaccessible markets—to consumers who had never interacted with those products in the past.

When you begin to interact with historically underrepresented consumers, new opportunities become apparent for growing your business. So, not only are these inclusive groups saving time and resources by highlighting limitations from the onset, but they are also expanding opportunities.

I encourage you to take one small step toward product inclusion. Commit to integrating at least one inclusive practice throughout your product design process—from ideation to testing. For example, conduct a focus group across two dimensions of diversity at the beginning and middle of your product design process and conduct inclusive user testing near the end.

When you consult with your target users both before and after you have a prototype or a viable product, you'll still be able to pivot based on their feedback, but will likely mitigate potential challenges or concerns much earlier, thus creating a more inclusive product at a lower cost in terms of both time and money.

Case Study: Grasshopper

Grasshopper is a learn-to-code application for adults with no prior computer programming experience. Grasshopper's mission is to bring more people into the field of computer programming, so the product team wanted to create an application for everyone, regardless of age, sex, socioeconomic status, or other differences. To achieve this goal, the product team adopted the following product inclusion best practices:

- Clearly defined the "why" behind their work—to enable *anyone* to learn to write code
- Thought about the people who historically had been underrepresented in computer science and intentionally built the product for them
- Integrated inclusive design throughout the process—from ideation through testing
- Solicited feedback from people of color, women, and other underrepresented groups in our inclusion champions dogfooding pool as they were building lessons or changing functionality in the app
- Made the app easy and fun to use, thus inviting more people to the platform

Grasshopper first launched on Android and iOS in 2018, and within its first month of public availability was downloaded and used by more than one million people. It has since received broad praise for its usability and functionality. According to one user, "Grasshopper shows me that no matter what or who or how I look, anyone can learn to code. It opens up a whole new world for me."

Here are a few takeaway lessons from this case study:

- Look for areas of exclusion to identify the underrepresented users you want to build for.

- Align the "how" of your work with the "why." When your "why" is to serve historically underrepresented users, your "how" must involve those users in the process.
- Integrate inclusive design principles and practices throughout the process—from ideation through testing.
- Design products that are fun and engaging as much as possible to build an even more inclusive consumer base.

Case Study: Google Duo

Google Duo is a video chat and messaging app that enables face-to-face calls with up to eight people. The consumer communications team that created Google Duo, Google Fi, and more has been committed to building inclusively across their product set. Dimitri Proano, Simon Arscott, and Stephanie Boudreau have led the product inclusion working group for the past few years, implementing concrete accountability measures to build for everyone, with everyone. In addition to the core product inclusion working group, various Googlers in the consumer communications team have stepped up to ensure that products are built with inclusion in mind.

With recruitment help from Google's product inclusion team, in coordination with Senior User Experience Program Manager Josh Furman, the Google Duo team was testing the application's low-light feature with Googlers of various skin tones. During testing, they discovered that this feature, meant to boost video call clarity in low light, was accidentally triggering incorrectly for some users with darker skin.

Because the team had accountability frameworks in place, such as objectives and key results (OKRs) around product inclusion, and a product inclusion working group within their team, they were able to bring in the right people to help solve their problem. Niklas Blum and Conor Steckler used captured videos from the product inclusion community to adjust the low-light detection mechanism and the sensitivity of the post-processing software. These tunings helped to prevent false triggering while also improving camera exposure for darker skin tones. The team also enabled users to toggle the feature on and off easily during a call. Ultimately the team is excited about launching a feature that improves the video calling experience for everyone.

What is most notable in this case study is that the Duo team was able to identify a problem prior to launch because it was testing the application with people of various skin tones. And because they already had a product inclusion working group on their team, everyone knew where to go for help when the problem was identified. With help from our inclusion champions, the team was able to overcome the challenge and launch with functionality that was sensitive to user needs throughout the feature set.

For your team, identify a champion (or a team of champions) who is the go-to person or group when someone identifies a problem or an issue, so everyone knows who to call. Many people may play a role in solving inclusion challenges, but having a designated champion or specialist makes it easier to bring in the right people to collaborate on an issue.

Remember, diverse perspectives bring insights that you and your team may not have imagined. From minor issues that detract from the user experience to huge gaps for underrepresented groups, the solution is to listen to and test with historically underrepresented users. The Duo team leaned into the "universal" part of Google's mission to ensure testing across geography and across the skin tones to make their product as universally accessible and appealing as possible.

Case Study: Project DIVA

Project DIVA (DIVersely Assisted) is the brainchild of Googler and Software Engineer Lorenzo Caggioni, who was inspired to making products more accessible by his brother, Giovanni. In this section, Lorenzo tells the story behind Project DIVA.

My 21-year-old brother Giovanni loves to listen to music movies. But because he was born with congenital cataracts, Down syndrome, and West syndrome, he is nonverbal. This means he relies on family and friends to start or stop music or a movie.

Over the years, Giovanni has used everything from DVDs to tablets to YouTube to Chromecast to fill his entertainment needs. But as new voice-driven technologies started to emerge, they also came with different challenges that required him to be able to use his voice or a touchscreen.

That's when I decided to find a way to enable my brother to control access to his music and movies on voice-driven devices without any help. It was a way for me to give him some independence and autonomy to be able to use the same technology other family members use to accomplish the same task.

Working alongside my colleagues in the Milan Google office, I set up Project DIVA. The goal was to create a way to enable people like Giovanni to trigger commands to the Google Assistant without using their voice. We looked at many different scenarios and methodologies that people could use to trigger commands, such as pressing a big button with their chin or their foot, or with a bite. For several months we brainstormed different approaches and presented them at different accessibility and tech events to get feedback.

We had a bunch of ideas on paper that looked promising. But in order to turn those ideas into something real, we took part in an Alphabet-wide accessibility innovation challenge and built a prototype which went on to win the competition. We identified that many assistive buttons available on the market come with a 3.5mm jack, which is the kind many people have on their wired headphones. For our prototype, we created a box to connect those buttons and convert the signal coming from the button to a command sent to the Google Assistant.

To move from a prototype to a more robust solution, we started working with the team behind Google Assistant, and we are announcing DIVA at Google I/O 2019.

The real test, however, was giving this to Giovanni to try out. By touching the button with his hand, the signal is converted into a command sent to the Assistant. Now he can listen to music on the same devices and services our family and all his friends use, and his smile tells the best story.

Here are the steps Lorenzo and team took to ensure that DIVA worked for their target user:

- **Finding inspiration for inclusion in exclusion:** Lorenzo found inspiration for his mission in a single case of exclusion—his brother Giovanni. He then extended his mission to include other people who faced product usability challenges.
- **Getting feedback from the target demographic early and often:** By starting with one use case and then getting feedback from others, Lorenzo and team were able to help Giovanni and others use

technology independently. Being able to do things on our own, to leverage technology to enrich our lives, is something we can all get behind. Testing and receiving feedback allowed the team to do just that.

- **Attending industry events to get alternative perspectives:** Giovanni and team attended professional conferences and sought feedback, which is a very effective and cost-efficient way to test ideas with people who represent a diversity of perspectives.

Remember, you don't have to invest a ton of resources getting user feedback. You can obtain valuable feedback by attending regular meetups, engaging with others in co-working spaces, and participating in industry and professional events. Seek feedback from people outside your team wherever and however possible. These are also great ways to test a hypothesis you have or to solve a key challenge for a certain demographic before you get too far along in the product design process.

CHAPTER 10

Making Your Marketing More Inclusive

Our definition of product inclusion covers everything from ideation through marketing. After all, the success of a product or service hinges on the organization's ability to get that product into the hands of as many consumers as possible, regardless of their differences and you do that through authentic connection and value.

Marketing plays an important role in the success of a product or service. If your marketing efforts fail to resonate with historically underrepresented consumers, your products will pay the price in terms of negative word-of-mouth, lost sales, and limited distribution. Organizations have a choice: they can make their marketing inclusive by representing the diversity of users or make it exclusive by focusing their efforts on a single demographic.

Marketing should be the culmination of all the previous chapters' work—an extension of the inclusive practices integrated throughout ideation, user experience design, and user testing. It should reflect the diversity of perspectives baked into the product; it should *not* be an inclusive advertisement for an exclusive product.

What makes marketing resonate? Three factors:

- **People:** If your organization has diverse representation across its brand, more people will feel included and welcome and open to buying or interacting with your product.

- **Product:** The product must reflect the care taken to accommodate the differences among individuals. People from underrepresented communities will readily spot any inauthenticity in the marketing of a product that wasn't designed with them in mind.
- **Story:** The story is key. People want to feel connected; it's a core part of the human experience. At the center is human connection; the product facilitates and enriches that connection.

At the end of the day, it's about the story. It's about your product. And it's about the people. Fusing these three components together and being inclusive in doing so is where marketing magic happens.

In this chapter, I cover key principles that drive our inclusive marketing at Google and present some examples across a variety of marketing media.

Inclusive Marketing Guidelines

At Google, we build products for billions of users around the world, and we feel a deep responsibility to ensure that everyone sees themselves positively represented in the stories we tell. Our approach to inclusive marketing is guided by this mission. We cast a net for our stories as wide as our user base, we tell stories that feel real to people and demonstrate our understanding that most of them have backgrounds and experiences much different from our own.

Our marketing teams don't see this as "multicultural marketing." For us, it's marketing in a multicultural world. As our Chief Marketing Officer, Lorraine Twohill, has said, "We don't have all the answers. We are on a journey of becoming inclusive marketers, and we have a long road ahead of us." In an effort to move us all forward together on this journey, Lorraine has shared the following learnings that guide our work:

- **Build teams that reflect the world around you.** This practice embeds diverse perspectives at the outset and throughout the creative process. Even so, everyone on the team is responsible for getting it right and for making inclusion an integral part of their thinking. In parallel, work with partners and agencies who also care about inclusion and who can bring diverse perspectives to the work. Foster an environment of psychological safety, so that everyone can speak freely.

- **Hero diverse stories.** Challenge yourself and your team to hero a different group and experience in every piece of marketing. Consider the body of work that you're contributing to, and embrace the opportunity to make it more representative of the multi-dimensionality of your users and their experiences.
- **Strive for authentic, relatable storytelling.** Tell real stories as much as possible. If you fictionalize, base it in deep user and cultural insight. Your mission is to tell stories that ring true with all users. Authenticity and relatability are key, so lean on insights and testing to help you get it right.
- **Challenge stereotypes.** Pursue positive portrayals that directly challenge common preconceptions and stereotypes. As Lorraine puts it, "No more women in the kitchen, please!" Bring diverse perspectives to the table that challenge stereotypes. Then, move beyond stereotypes to authentic, multidimensional portrayals. One of the best ways to do this is to use real footage filmed by everyday people.
- **Think beyond casting.** Casting underrepresented groups isn't enough. Make sure every aspect of the creative represents their experience. Consider story, setting, music, voiceover, family dynamics, wardrobe, food, product portrayal, etc. And don't forget about the people behind the camera: the editors, producers, directors, and collaborators. They often make all the difference.
- **Understand your brand's role.** Know the cultural and brand context around those whose stories you're telling, and make sure the brand can credibly and positively participate in the conversation. Inclusive marketing is a fundamental brand commitment, not a PR opportunity, so the brand shouldn't be the hero.
- **Hold yourself accountable.** Track and audit your work over time to understand where you are making progress and where you may be falling short. Google uses a combination of machine learning and manual reviews to track progress. This approach has uncovered new learnings that continue to shape how our teams think about inclusion. For example, a recent audit showed 23 percent of the people in our U.S. ads were Black but that we were casting a disproportionate number of interracial couples, mostly light-skinned individuals, and putting people in stereotypical roles, such as dancing and playing music and sports. Having that insight helped us develop a plan to improve going forward.

These guidelines help to frame how all marketers should think about and approach their work.

Google's Inclusive Marketing Consultants Task Force

During my time on the product inclusion team, I've had the opportunity to collaborate on the creation of our inclusive marketing consultant's task force: a 60-person (and growing!) group of marketers from underrepresented backgrounds who review many of our marketing team's campaigns with an eye toward inclusion—before they are released into the world. This group, executive-sponsored by Sherice Torres, a Director in the Google Brand Studio, works to lead teams in the right direction and to proactively prevent bias from creeping into our marketing campaigns.

Based on the initial results, we have found that getting involved early and often leads to creative that more accurately represents the world around us. Now, we are working hard to scale these reviews across all of our campaigns.

Lorraine's team has invested considerable time, effort, and expertise to create inclusive principles and practices. I've been fortunate enough to work with and learn from Brand Marketing Manager Raphael Diallo, a member of Lorraine's team who leads inclusive practices across the marketing organization. Raphael is tasked with creating actionable practices and methods of implementation across the entire organization. He is the embodiment of integrity and exudes passion for the inclusive principles that drive his team's work (see the nearby sidebar).

We have the world's attention. Let's use it wisely

—Raphael Diallo, Brand Marketing Manager

"You might watch a movie once, but you'll see an ad 40 times."
—Eva Longoria at ADCOLOR 2019

Underrepresentation and stereotypes are harmful. They can lead to unconscious bias, active prejudice, and low self-confidence among children and adolescents.

We have a responsibility to positively contribute to the media landscape.

Personally, this is why I care so much about inclusion in advertising. Every brand should be pushing hard to make sure the creative we build is, bit by bit, making the world a better place. That means taking inclusion into consideration with every creative choice we make. Because if we don't, we will unintentionally exclude our users.

Several years ago these insights were not top of mind. Our team often saw anecdotal examples of our marketing coming up short. For example, the lack of darker skin tones in product photography or events that didn't have gender parity on stage. But, it wasn't until we did an extensive audit of representation and portrayal in our creative did we realize the extent to which we needed to improve.

This data prompted us to build resources and structures to help tackle the problem. We rolled out research-backed guidelines, workshops, and an Inclusive Marketing Consultants task force for our marketers. We also launched an internal awareness campaign to drive Googlers' engagement with these resources.

We're continuing to measure our progress through regular audits of our creative. Ultimately, we can have high engagement with the guidelines and workshops—but if the results of those efforts don't show up in our work, we won't be successful.

Inclusive marketing is not a simple box to check. But it is an opportunity to reach new users—and deepen relationships with existing ones—through more relevant channels and authentic creative. Along the way, we also have a chance to make a positive contribution to the media landscape by eliminating harmful stereotypes and portraying the historically underrepresented.

Extending inclusiveness to all your marketing activities

As you integrate product inclusion into your marketing, extend it to all your marketing activities. You want an inclusive mind-set to permeate everything you do, from posting blog content, to sponsoring events, to developing global campaigns.

Keep in mind that authenticity is crucial, and that begins with diversity and inclusion across your organization. If you truly care about people in underrepresented communities, your organization will provide opportunities to hire and partner with members of these communities and share in their success.

Whether you run events, make commercials, or are part of the team creating your brand's narrative, bringing diverse perspectives to the forefront creates a strong, cohesive, story that all users can and want to get behind.

Taking a global perspective

—Maria Clara, Head of Search Brand Marketing, Google

The quote from Joe Gerstandt found at the beginning of this book says a lot about the work we started in recent years with different Google marketing teams in Brazil. We were aware that it was necessary to take some important steps to increase representation of the Brazilian population in our communication. This has been made even more clear by data from recent research, which shows the opportunity in front of marketers to produce advertising campaigns that effectively include diversity.

If we invest in making our products accessible to as many people as possible, we understand that the external communication associated with these products has a fundamental role. When we use our communication to give visibility to certain groups, we can help uplift voices that have historically been marginalized. And that is where the strength of brands is: thinking about campaigns that provoke reflections on the status quo, that break stereotypes and bring a positive light to harmful patterns in our society. But where does this journey begin?

In a company the size of Google, there we want to be sure that we adhere to global values while looking at the local reality.

Our first step was to identify internal opportunities as it related to team composition. One that stood out was that we wanted to ensure we had more Black women's voices represented. We knew that we would need to hear from experts, from academics to influencers (both externally and internally, including the AfroGooglers committee, an internal employee resource group). We wanted to learn to see beyond our own unconscious biases. So we used research methodologies to understand the extent of this need to be more inclusive within the context of our country. And finally the third step was to engage the people involved, from our team and from outside, in an open dialogue.

We did research work with Black women to understand this reality within our area of expertise. In addition to our own films, more than 70 advertisements were evaluated—from different industries—in which there was an intentional movement by brands to include Black women in their narratives.

The diagnosis was very clear: the Black women we interviewed said that they did not feel represented by any brand today, with the exception of some beauty brands that, in their view, had already understood the importance of correcting this route for commercial reasons.

We talked about the challenges of implementing global values locally in a company the size of Google, but it's also important to mention the numerous benefits. The biggest one here, perhaps, is that we are inserted in a giant ecosystem that generates and exchanges learning at an incredible speed. And that exchange proved to be a key point to promote diversity and inclusion initiatives.

Findings:

As a result of our work,

Google has grown +12 ppts in awareness, +6ppts in admiration, and +7ppts in advocacy among lack population (H2'18 vs H1'19)

Google is increasingly perceived as "a brand that cares and portrays different people and culture" +9ppts, "without reinforcing stereotypes" +6ppts.

No brand will be able to handle everything alone. Without involving and learning from our suppliers and partners, it is not possible to generate demands in the market so that it can actually transform.

As you integrate inclusive design into your marketing, extend it to all your marketing activities. You want an inclusive mind-set to permeate everything you do, from posting blog content, to sponsoring events, to developing global campaigns.

Keep in mind that authenticity is crucial, and that begins with diversity and inclusion across your organization. If you truly care about people in underrepresented communities, your organization will provide opportunities to hire and partner with members of these communities and share in their success.

Whether you run events, make commercials, or are part of the team creating your brand's narrative, bringing diverse perspectives to the forefront creates a strong, cohesive, story that all users can and want to get behind.

CHAPTER 11

Measuring Product Inclusion Performance

When you begin to integrate product inclusion into your work, you need to find ways to gauge its impact, so you can measure your team's performance and its progress toward meeting its objectives. In business, you gauge performance through the use of *metrics*—quantifiable measurements used to track, monitor, and evaluate success and identify areas needing improvement.

One challenge that teams often encounter when getting started with product inclusion is deciding or agreeing upon which metrics matter. A number of metrics are available for measuring success and failure across the product design and development process. Some points in the process are more difficult to measure, and what you choose to measure may evolve over time, but initially, your team needs to establish an initial set of metrics to serve as a starting point.

In this chapter, I introduce a few key metrics for measuring your product inclusion performance or progress, which you can use right out of the box or as inspiration to start thinking about the mechanism your team or organization wants to use to measure and track its progress. Closely monitoring and analyzing the metrics you decide to use ensures that your team/organization stays on track, and it provides the motivation that drives everyone toward achieving the stated objectives.

Mastering Performance Measurement Fundamentals

Measuring performance is about more than just having a set of metrics in place. You need to be sure the metrics align with your objectives, that everyone on your team knows what those metrics are, and that you are continuously taking measurements, tracking them, and using them to make adjustments. To ensure that you are optimizing the use of metrics to drive continuous improvement, follow these suggestions:

- Define your objectives first (see Chapter 6). Metrics are closely tied to objectives, so before choosing metrics, look at what your team or organization is trying to accomplish in terms of product inclusion, such as increasing leadership commitment or employee involvement by a certain number or percentage or making a certain product more usable and appealing to users from a certain demographic.
- Choose metrics that provide a quantifiable measurement of performance or of progress toward meeting the specified objective. Quantifiable that the metric must be a numerical value.
- Be sure that each metric is clear and actionable, meaning steps can be taken to improve results if the metric shows that improvement is necessary. Measuring something you cannot change just because you find the metric interesting is not helpful. An example of an actionable metric would be a certain percentage increase in user satisfaction. A less actionable metric would be something like 100,000 new visitors to our website if the goal is to have people purchase a product. Instead, you would want to know how long visitors stayed on the site, the bounce rate, the abandoned-cart rate, and so on—metrics that would provide insight about *why* people visited the site and purchased or did not purchase a product.
- Make sure everyone on the team is aware of the objectives and metrics. When people know how their work is being evaluated and they can see progress, they become more empowered and encouraged to do the work, to eliminate what does not work, and to be proactive about applying an inclusive lens to what they do on a daily basis.
- For each metric, identify a method for taking the measurement. Ideally, data collection is automated, so it does not distract from the

activities being measured, but some metrics may require that people in the organization report data or complete and submit survey forms.

- Set a frequency for taking measurements and for examining results; for example, weekly, monthly, or quarterly. For our product inclusion team, I like to pull most metrics at least once a quarter, so we have time to pivot before the date set for meeting our goal, but this will vary according to the work being done and the objective.
- Take measurements prior to implementing an initiative or integrating product inclusion in a process to establish a baseline for comparisons; otherwise, you will have no way of gauging the impact of changes on outcomes.
- Determine who will be collecting data, analyzing that data, and reporting back to the team. Does a central analyst group collect and analyze metrics for the entire organization? Is each team responsible for its own metrics? Is a particular individual on each team in charge? If you have both central and team metrics, find a way to bring them together to provide for more cohesive reporting.

Following these suggestions optimizes the use of metrics in driving positive change and facilitates the process of empowering and motivating people to get on board. The proper use of the right metrics enables everyone to focus on clearly defined objectives, see the progress being made, and identify the impact of their decisions and actions on the outcomes. Metrics are great tools for encouraging and engaging people by showing them that the work they do matters and that it is contributing to the success of the organization and the customers being served.

Choosing the Right Performance Metrics

Good teams have "metrics that matter." Great teams include metrics that focus on underrepresented users, and they ensure that every metric leads back to those users' needs and preferences and to the organization's business goals.

In this section, I encourage you to start with metrics you may already be using and adapt them, wherever possible, to measure performance in relation to your organization's or team's product inclusion objectives. I then present several metrics our team uses at Google to measure and track our product inclusion performance and that of other teams.

Evaluating your metrics needs

Before you start adding product inclusion metrics for your team or organization, examine the metrics you already have in place. You may be able to use these metrics as is or modify them to bring them more in line with your product inclusion objectives. As you examine the metrics you are already tracking, ask the following questions to determine whether some existing metrics can be modified and to identify where additional metrics may need to be added:

- What metrics do you currently use to assess the performance of your product or service, the impact of changes in the process or practices used to build or improve that product or service, or the progress of your product inclusion initiatives?
- Do your existing metrics provide insight into the needs and sentiments of non-majority users specifically? (You probably have metrics to track user needs and sentiment, but make sure they extend specifically to the needs of underrepresented users.)
- Can you see a way to apply an inclusive lens to any of your current metrics? For example, if you already track user sentiment based on socioeconomic status, you could extend that metric by geography to evaluate what users in areas of rapid growth think about your product, or you could expand the metric to include users who earn below a certain bracket you currently are not tracking.
- What does "underrepresented user" mean for your organization, team, process, or product? The meaning may differ based on industry, product, demographic, or other factors. In technology, for example, an unrepresented user may be someone who is just beginning, whereas in the food industry, an unrepresented user may be someone with certain food intolerances or sensitivities, or someone who may not have access to certain foods regularly.
- Where do you *not* have insight into the needs and sentiments of underrepresented users? If you do not even have a clear idea of who your underrepresented users are, consider conducting focus groups that span multiple demographics. You can then conduct additional focus groups with people representing a narrower set of demographics to home in on their needs and sentiments.

During this process of evaluating metrics, modifying existing metrics, and deciding which metrics to use, be sure you have consensus among all stakeholders and decision-makers regarding the final set of metrics you settle on. One great way to gather input and achieve consensus in a relatively short period of time is to run design sprints specifically for discussing inclusion metrics. See Chapter 8 for more about running product design sprints.

Classifying metrics

When deciding which metrics to use, thinking about different classifications or groups of metrics can be helpful. Our team breaks down metrics into different buckets to better understand them in terms of their application or function. One way we classify metrics is to divide them into two buckets we refer to as socialization metrics and product inclusion (PI) metrics:

- **Socialization metrics** are measurements we use to track progress in terms of diversity and inclusion awareness and participation across the organization. Here are a few examples of socialization metrics:
  - Number of leaders engaged in product inclusion
  - Number of product areas or business units that have diversity and inclusion objectives and key results (OKRs)
  - Number of volunteers working to support product inclusion
- **Product inclusion (PI) metrics** are measurements used to track progress in terms of integrating product inclusion into what our product teams are doing and the outcomes of those efforts. Here are a few examples of product inclusion metrics:
  - Diversity of representation on a team
  - Number of users buying or engaging with a product or service
  - Number or frequency of negative user experience reports/escalations

Another way we classify metrics is by dividing them into input and output metrics:

- **Input metrics** track the resources needed to produce a desired outcome; for example, 300 volunteers attended last month's event.

- **Output metrics** reflect the outcome resulting from a given input; for example, one month after the event, use of our product inclusion dashboard increased among people in our organization by 30 percent.

Note that one team's output metric may be another team's input metric. For example, part of our product inclusion team's focus is on broader socialization objectives—increasing awareness and adoption of diversity and inclusion across the organization. For us, an input metric may be the number of meetings we have with various teams in a given quarter, and the output metric may be the movement of representation throughout the product design process by bringing multiple voices through our various programs. For a product team, on the other hand, the diversity of representation on the team may be an input metric, with the output metric being the number of innovations incorporated into the new version of the product or the product's sales volume in the first three months after product release.

Considering some metrics to adopt

In this section, I present several metrics that organizations, business units, product areas, or product teams may want to consider using to track their performance and progress. Every team should track both input and output metrics to gauge the impact that changes in inputs (people, processes, and practices) have on their outputs (product functionality, user engagement, negative feedback, and so on).

Input metrics Here are several input metrics that business units, product areas, or product teams may want to use to track the changes they make to their people, processes, or practices in an attempt to improve the inclusivity of their products or services:

- **Number of leaders engaged in product inclusion:** By "engaged," we mean talking to their teams and peers, establishing objectives, defining metrics, developing inclusive processes, and talking to people who represent diverse perspectives.
- **Number of product areas or business units that have diversity and inclusion objectives and key results (OKRs):** At Google, all

product areas and business units use OKRs, so if they have an OKR around product inclusion, this means they've taken the first step to prioritizing this work as core to the way they do business.

- **Number of employees per product area, business unit, or team mobilized in service to product inclusion:** The number of people in itself is not an indicator of progress, but how the number changes over time is. A large increase may represent growing momentum and a groundswell of people who think this work is important and are committed to moving it forward.
- **Team representation:** The diversity of team members has an impact on the inclusiveness of the products the team creates, so keeping an eye on the diversity of representation on each team (especially on product, research, and marketing teams) provides valuable insight on how changes to this metric impacts the product and user engagement and sentiment regarding that product.
- **Number of volunteers working to support product inclusion:** The number of volunteers that support product inclusion and invest their time and expertise in ideation, design, and testing can have a significant impact on making products more inclusive.
- **Budget:** The amount of money invested in training and tools, and in exposing leaders and employees to underrepresented groups (such as trips to foreign countries) may be a predictor of the success of product inclusion initiatives.

Output metrics Output metrics are used to evaluate the outcomes that result from changes in inputs. These are the metrics that generally attract the most attention because they show whether your team is moving in a positive direction, to what degree, and at what pace. Output metrics may vary by industry and product, but generally include the following:

- **User engagement or total number of users:** Presumably, if you are changing inputs to improve a product and make it more appealing to a broader consumer base, the number of people who buy or use your product should increase, especially among the underrepresented populations you are trying to serve.
- **Conversions:** Conversion rates can represent the intersection of efforts to be more diverse and inclusive across the entire product design,

development, testing, and marketing processes. Several factors may contribute, alone or together, to move this number up or down.

- **Customer satisfaction/brand loyalty:** Improvements in products should lead to an increase in customer satisfaction and brand loyalty, especially in products or marketing that demonstrate the organization's sensitivity to the needs of diverse users.

- **Number or frequency of negative user experience reports/ escalations:** An increase in number of users coupled with a decrease in the number or frequency of negative user feedback is a good indication that product quality and appeal has improved for a more diverse population. This metric can be applied to content, as well, indicating (through reader feedback) whether readers are finding content to be more or less offensive, for example.

- **Number and quality of innovations, new features, or new functionality related to diversity/inclusion:** Increasing diversity and inclusion throughout the product design and development process should result in an increase in the number and quality of innovative features or functionality specifically added to serve unmet needs of historically underrepresented populations.

Recognizing the importance of employee satisfaction

—Tomas Flier, Community Advisor of the Latinx ERG and former Product Inclusion Analytics Lead who helped develop our metrics

The message that product inclusion is sending to our communities is about the need of businesses to have us in their teams. Our unique backgrounds make our input extremely valuable for those businesses to be successful. When we are talking about inclusion, the two main pillars that we look at are not only how the different groups feel they belong but also how they feel valued, not for being similar to others, but on the contrary—for being valued for their uniqueness. As the community advisor of the Latinx Employee Resource Group (ERG) I can see how these Googlers from underrepresented communities feel included when they begin to understand how their differences are now being valued through the contributions they make as a result of their unique perspectives. Their cultural differences now make them unique and extremely valued.

On top of that, we want products that are relevant for us. I want a smart speaker that can understand my accent and can answer all the questions about soccer I have every day!

The Google Assistant team actually incorporated product inclusion before launching it. They were focused on building inclusively from the very beginning. Today, the Latinx community represents around 18 percent of the U.S. population and accounts for half of the U.S. population growth. They represent 74 percent of the workforce growth in the U.S. On top of that, the early adopters of these new technologies are usually young audiences, and more than half of Latinos are under 33 years old. Every 30 seconds, two non–Hispanics hit retirement age, and one Hispanic turns 18 years old.[1]

Due to the importance of this community, the Google Assistant Team included many Hispanic/Latinx voices throughout the product development process, especially in the testing phase. The result ended up being a much more inclusive product, not only because it includes Spanish as an option but also due to its capacity to process different accents. Today, the Google Assistant understands me very well.

Combining Metrics with Objectives and Timelines

It has[1] been said that a goal without a deadline is just a dream. I recommend that for every metric you use, you tie it to an objective and a timeline complete with milestones to keep you and your team focused and on track. Consider creating a table like Table 11.1.

Our team created this table (only a portion of which is shown in Table 11.1) to provide us with a quick reference tool that could always drive us back to what mattered and what we had all agreed upon. This

---

[1] Pew Research.

Table 11.1   Tie metrics to objectives and timelines.

Goal/Objective	Metric	Timeline
XX percent increase leadership shift from awareness to action	Number of leaders with a product inclusion accountability framework (like OKR) in place	Accountability frameworks due by Q2 for commitment to action by year end
XX percent increase employee sentiment as it relates to product design practices	% of employees who provide positive reactions to product inclusion work	Measure and establish baseline by end of Q2, measure every six months with objective deadline of year end
Increase employee allyship with historically underrepresented users by XX percent	Number of employees with at least one actionable commitment to product inclusion	Implement phase 1 by end of Q2 and phase 2 by year end, measure quarterly after implementation with objective deadline of end of Q2 the following year
Increase representation of underrepresented team members on XX percent of product teams	Number of team members and volunteers (who reflect underrepresented users) involved with product teams on product design, development, and testing	Measure and establish baseline by end of Q4 with objective deadline of Q4 the following year

table serves as our navigational North Star, keeping us from getting nudged off course by other metrics and distractions.

Checking these metrics regularly, we have observed a continuous increase in engagement, both from the top down and the bottom up (from leadership and Googlers) across the company. By coupling these metrics with some of the tactics to get buy-in described in Chapter 4, you will begin to see momentum build as a result of your efforts. You'll also have more clarity into what you measure and why. You'll be able to segment and understand feedback more granularly, which is something that all businesses want because it enables you to make more targeted changes and trace your organization's or team's path to success.

CHAPTER 12

The Many Shades of Nude: Product Inclusion in Fashion and Retail

Imagine how people of color felt when all the "flesh" colored bandages on the market were light pink. What is this saying to certain consumers? Is their flesh not flesh? Imagine how a person must feel when the XL or plus sizes a store carries are too small. Think how someone in a wheelchair must feel when he goes shopping and cannot reach half the items on the shelves without having to ask for assistance.

If you cannot imagine the frustration an underrepresented consumer may feel in the retail space, think about a situation in which you've gone to a website on your phone and it wasn't optimized for mobile use. The text is too tiny to read, and when you zoom in, you can see only a small portion of the page. You try to rotate the phone, and the page header takes up so much screen space you can't see anything else. Exasperated, you may have given up and left the site or headed off to look for a site with similar content. The experience is certainly different, but the feeling of exclusion and frustration over being denied access to something you need or want is not a positive one.

Consumers who do not conform to the parameters established by the majority of the population are often forced to feel as though they have been condemned to live in a world that was not designed for them. They may be like most of the population in every other way, but due to one dimension of diversity that singles them out, they are made to feel unwelcome and ignored.

In order to demonstrate that product inclusion is not only a tech-specific challenge, in this chapter, I focus on what can be done in fashion and retail businesses specifically to make products and shopping experiences more inclusive, and I present examples of how Gap Inc. and designer Chris Bevans have prioritized inclusion to do the right thing to their own competitive advantage.

Focusing on Inclusive Fashion

Fashion is a form of self-expression and is often integral to a person's identity. It can be used to blend in, stand out, or make a statement about who you are, what you believe, and with whom you associate. Fashion is also one of the first things you may notice about a person. It serves as a window into who people are and how they view themselves.

Fashion companies that lean into the responsibility to provide options for everyone have much to gain, and those that do not have much to lose. In 2014, *Fashion Spot* (a fashion industry forum) called out several fashion magazines for a lack of diversity on their covers.[1] One magazine had gone 12 years without featuring a model of color on its cover.

Fashion magazines have a long history of discrimination against models of color based on the mistaken notion that Black models don't sell. Evidence proves otherwise: "From Lupita Nyong'o on American *Vogue* to Kerry Washington on *Vanity Fair*, women of color have earned positive responses from readers, not to mention have shown no measurable impact on sales."[2] Winnie Harlow is another great example of a beautiful model from an underrepresented group (people with the skin condition vitiligo) who is very popular and successful.

---

[1] https://www.thefashionspot.com/runway-news/509671-diversity-report-fashion-magazines-2014/.

[2] https://www.mic.com/articles/107564/one-fashion-magazine-just-ended-12-years-of-exclusion-in-a-beautiful-way.

As diversity and inclusion become mainstream in fashion, more and more companies will be embarrassed into doing the right thing or will simply lose business for not doing so.

Considering color

In the fashion industry, color plays a major role in product inclusion because it can be very alienating. If you can't find the right tone for makeup, for example, you may have to purchase multiple products to blend to the right shade or forgo that brand or product all together. This makes a customer feel "othered" and it can be a deeply painful experience. We should all be able to walk into a store, find our perfect shade, and not be made to feel that "nude" doesn't apply to us.

Cosmetics are not the only items to consider. When you think about anything from pointe shoes to ballet slippers, tank tops, underwear, bandages and more, it can be very alienating to see time and time again that the color of "flesh" or "nude" (the default color) is nothing like your own. What message does this default send to consumers of color: Do they not have flesh? Are their nude bodies discolored? The further away someone is from that default, the more dehumanizing it can feel. There are stories of people who cried when finding a bandage that matched their skin tone.

Broadening selections for size and fit

For decades, the fashion industry has been producing and selling clothing based on its definition of ideal. The "sample size" as defined by the industry is 5 feet 10 inches tall, 115 pounds, size 2–4. Even though the average American woman is 5 feet 4 inches tall, 164 pounds, and size 14, sizes at many stores top out at 12. Even though the average American man is 5 feet 9 inches tall and 195 pounds with a 40-inch waist, in many stores the top waist size is 38 inches. To compound the problem for both men and women, sizes often fail to account for differences in shape; two people, each weighing 170 pounds can be shaped vary differently due to genetics, muscle tone, and other factors. There are brands that have intentionally started to highlight and focus on inclusion as a core business practice, but like many industries, there is a lot of opportunity to create more inclusive products across multiple dimensions of diversity.

In footwear, the problem is also prevalent, with most shoes topping out at size 13 and for adults with small feet, it can be challenging to find shoes as well. To add insult to injury, sometimes there is a premium for the solution—larger sized clothing. On the opposite side of the spectrum, smaller women and men are also excluded, some having to choose clothing designed for children or teenagers.

Fortunately, many retailers are beginning to realize the disparity between the sizes they carry and the sizes and shapes of their shoppers. SmartGlamour, an independent New York city retailer, offers clothing for women of any size. As stated on its website, "Every design is available in XXS to 15X and beyond. Every item can be customized to fit any and every body." Other retailers are beginning to get the message and broadening the range of sizes they carry. Given that the plus market constitutes an estimated \$21 billion of the fashion sector, this is a wise business decision, and the right thing to do.

Making sensitive product and marketing choices

Although companies and brands don't intend to harm people, insensitive products and advertisements embarrass consumers and undermine the mission of the guilty parties—to serve consumers and grow sales. In addition, such incidents are easily avoidable. Designers and fashion companies should always have their concepts reviewed by a diverse group of people for feedback prior to release.

In addition, fashion companies need to be intentional in their marketing to ensure consistency with the company's stated commitment to diversity and inclusion. Chrissy Rutherford, Special Projects Director of Talent and Social at Bazzar.com, the e-magazine version of *Harper's Bazaar*, once had a brand reach out to her via social media for a potential collaboration. After reviewing the brand's Instagram posts, Chrissy advised the brand, "You have to put your money where your mouth is. You haven't Instagrammed a POC [person of color] since last summer, and I find that really concerning."[3] People will see through having inclusive marketing and a non-inclusive product.

---

[3]https://www.documentjournal.com/2019/02/the-cfda-addresses-why-we-still-need-to-talk-about-diversity-and-inclusion-in-the-fashion-industry/.

Increasing gender diversity

Given the fact that women spend an average of three times more on clothing than do men,[4] you might think that the industry would be run primarily by women. However, only 14 percent of major brands have a woman in charge according to a report published by the Council of Fashion Designers of America, Inc. in partnership with *Glamour* magazine.

The Council of Fashion Designers of America (CFDA is a not-for-profit trade association, founded in 1962, whose membership consists of nearly 500 of America's foremost womenswear, menswear, jewelry, and accessory designers). Google's product inclusion team has collaborated with CFDA to promote product inclusion in the fashion industry, which originally started with former Googler Jamie Rosenstein Wittman and continued on with myself and Jackson Georges, Vice President at CapitalG.

After the Women's March (January 21, 2017), CFDA and *Glamour* magazine conducted a study to examine women's empowerment and equality in the workplace, specifically in the fashion industry. Results from the study were published in 2018 in a report called "The Glass Runway: Gender Equality in the Fashion Industry."[5] Here are some of the highlights from the study:

- Gender diverse companies outperform companies dominated by one gender.
- While 100 percent of women surveyed see gender equality as an issue in fashion, less than 50 percent of men do.
- At the vice president level, women begin asking less and receiving fewer promotions.
- Women have greater difficulty than men in juggling parental responsibilities, especially at the vice president level.

The study identified four clear actions that fashion companies can take to address gender inequality:

- Develop a compelling business case for gender diversity.
- Increase transparency and clarity of evaluations, promotions, and compensation.

---

[4]https://www.mckinsey.com/industries/retail/our-insights/shattering-the-glass-runway.
[5]https://cfda.imgix.net/2018/05/Glass-Runway_Data-Deck_Final_May-2018_0.pdf.

- Offer sponsorship programs geared toward empowering women.
- Create programs and policies that give employees the flexibility to fit work into their lives.

Given the fact that women spend more money on fashion and no doubt have perspectives based on their own experiences, fashion companies would be wise to increase gender diversity. In addition, women's brands have a harder time getting financing or may not be getting as much visibility, so supporting these brands is important to expand the ecosystem.

Making Your Retail Store More Inclusive and Accessible

In the retail business, you may think that product inclusion does not apply to you, but it certainly does—perhaps twice as much. Product inclusion applies both to the products you sell and to the experience you deliver via your store. Thinking about what the experience is like for underrepresented users is crucial to serving the needs and preferences of the greatest number of people.

In the following sections, I cover some of the considerations to make to ensure that your store and the products you carry are accessible and inclusive, whether you run a brick-and-mortar or online store.

Brickandmortar establishments

There's a palpable sense of warmth and eager expectancy when you walk into a store that makes you feel welcome and appreciated. It puts you at ease knowing that you are about to engage in a pleasurable shopping experience. This feeling can be amplified among those from underrepresented groups who are unaccustomed to experiencing consideration in the retail space. The sense of being welcome and included can melt away any fears or anxieties over the possibility of exclusion.

Creating the right atmosphere in a store does not happen by mistake or luck. It requires careful attention to detail and to factors you may not have thought much about. Here are several factors to consider:

- **Accessibility:** Complying with Americans with Disabilities Act (ADA) requirements is just the start. Here are some additional steps you may want to take:
  - Place heavier items on lower shelves and lighter items up high when possible.
  - Avoid placing displays in the middle of an aisle, which could obstruct movement for people on crutches, in wheelchairs, or people with vision impairments.
  - Prepare clerks to assist customers with removing products from shelves, but don't assume without asking or being asked to help.
- **Staff:** Hire a diverse set of employees. In addition to demonstrating your commitment to inclusion, you will reap the benefit of having a diverse staff's input on how to make your store more attractive to people from different communities. You also want to have diversity and inclusion training for all staff members—having a representative staff is a good start, but all employees need to think and act inclusively.
- **Security:** Security may be necessary to protect the safety of your staff and customers and prevent theft, but if you have security guards, they should not make people feel unwelcome or suspected. (Note that this disproportionately happens to people of color.) Here are a couple of suggestions to prevent common issues:
  - Provide bias training for security guards.
  - Poll your customers to determine how the presence of the security guards makes them feel and make adjustment as necessary.
- **Context and cultural nuance:** Make sure your store doesn't include anything that is likely to be perceived as offensive to a certain community. For example, if you have phrases in a different language, did you have people who speak that language natively work with you to create the design and ensure that it lands well?
- **In-store displays:** In-store displays, advertisements, and models (or mannequins) provide opportunities to make your store feel more inclusive. Here are a few ideas you may want to test depending on your market:
  - Group clothing by style, not size, eliminating any "Plus Size" sections.
  - Avoid genderizing sections like "toys for boys."

- Use genderless mannequins across all sections. If not possible, you can have mannequins typically found in the women's section dressed with what would be an outfit traditionally found in the men's section and vice versa, in order to be inclusive on not only styles but in gender identities. You could consider highlighting non-binary designers or designers who make gender neutral styles.
- Place a mannequin wearing the latest styles in a wheelchair or with assistive technology.
- Display models of different body types, hair textures, races, colors, ethnicities, and genders.

- **Cosmetics:** If you sell cosmetics, make sure the person choosing the color palette is sensitive to differences in skin tones and obtains feedback from people with different skin tones. If you have employees who help customers choose cosmetics, be sure they are trained to work with different skin tones. (I once went to a luxury makeup counter and as soon as the woman saw me, she went to get the one Black representative without saying a word, as she seemingly did not know how to support me in purchasing flattering shades.)
- **Hair care products and services:** If you sell hair products or provide hair cutting and styling services, carry a selection of products for different hair types and train your stylists to work with different hair types.
- **Clothing sizes:** Provide as much diversity in sizes as possible and consider making this your policy: same style, same price, regardless of size. Although manufacturers and retailers often justify charging more for plus sizes by citing the cost of additional fabric, that added cost is negligible compared to the costs of development, distribution, and retailing.

These are just a few of the areas to pay attention to. I strongly recommend that you test changes in your store with a diversity of shoppers to highlight other areas that may need attention. Prior to implementing a technique to make your store feel more inclusive, test it with the demographic you are trying to reach. Authenticity is crucial. If you introduce a change merely in an attempt to win an underrepresented community's business, it could come across as exploitive and actually hurt business. Remember, getting commentary and feedback from real customers is the only way to truly understand how an action or message will be received. Note that sometimes, best practices for one group actually helps everyone. Many of these best practices do.

Online stores

If you have an ecommerce store, pay attention to many of the same areas I highlighted in the previous section for brick-and-mortar establishments. For example, in the contextual images you use (people wearing or using your products), be sure to reflect the diversity of the customer base you are trying to reach. If you sell cosmetics, offer a comprehensive color palette. If you sell clothing, offer a broad range of sizes or custom sizing.

With online stores, accessibility may be the bigger challenge. Shoppers may be using a desktop, tablet, or smartphone. Some may have extremely slow Internet service. Others may be using a screen reader to accommodate for a visual impairment. Some may be navigating with a keyboard instead of a mouse or may have difficulty positioning the mouse pointer over certain objects on the screen. Some may be color blind and have difficulty distinguishing a button from its background depending on the color scheme. If your online store fails to account for these differences among shoppers, you will alienate some shoppers at the expense of your profitability.

Here are a few suggestions to ensure that your ecommerce store is accessible to a diversity of shoppers:

- Test the site yourself, experiencing it as a shopper would. Note that you will likely be more familiar with everything, but you may catch things that aren't as intuitive as they could be.
- Use an online accessibility tester to evaluate your site, such as WAVE (the Web Accessibility Evaluation Tool) or Google's Lighthouse extensions, and follow its recommendations and instructions to address any issues.
- Take a look at accessibility design guidelines at https://material.io/design/usability/accessibility.html#.
- Use a disability simulator, such as Vischeck (for color blindness) or a Designer (for visual impairment), to test your site.

- Make sure you can easily navigate your store with only a keyboard (no mouse).
- Include alternative (alt) text for all images.
- Add subtitles/closed captioning or a transcript (ideally both) to any product videos.
- Make sure you can easily navigate the store on a variety of devices—desktop, tablet, smartphone—and if you zoom in up to 150 percent or more.
- Compose product descriptions that include more than visual descriptions; for example, describe the texture or feel of clothing.
- Move the most important content closer to the top of the page to reduce the need to scroll down.
- Include an option for shoppers to call to place an order. (Some shoppers still feel uncomfortable entering account information online.)
- Test and optimize your site for speed to accommodate shoppers with slow Internet service.
- Solicit feedback about your store from ecommerce shoppers post sale, or use a service like UserTesting.com to have a diverse group of users test your site.

Product Inclusion at the Gap

My friend Bahja Johnson is a Director of Fashion Merchandising at Gap Inc.—a global retail and clothing brand. About a year and a half ago, she invited me to come to their offices in San Francisco to discuss what we were doing at Google as it related to product inclusion. I brought one of our original executive sponsors, Bradley Horowitz, to the meeting, as he is a product visionary and he pushes and questions in ways that make everyone think.

When we arrived at Gap's (beautiful) offices, we walked through the 101 of product inclusion with senior leaders who included:

- Mark Breitbard, Brand President & CEO, Banana Republic and Color Proud Council Executive Sponsor
- Michele Nyrop, Executive Vice President and Chief People Officer, Gap Inc., and Color Proud Council Executive Sponsor

- Margot Bonner, Senior Director, Gap Inc. Digital Marketing
- Maria Febre, Director, Global Diversity and Inclusion and Color Proud Council Corporate Sponsor
- Jermaine Younger, Senior Director of Merchandising, Gap Brand and Co-Lead of Color Proud Council

We talked about the business case for inclusion, including the similarities and differences between the retail/fashion and technology industries. Fashion and tech are interesting, in that even though their products are so very different, clear similarities exist, such as the diversity of users and user preferences; the movement toward customization and on-demand access; and the many users who have a strong affinity with specific brands and expect to build a relationship and trust.

During our visit, we learned a great deal about why product inclusion matters in the fashion and retail industries, and also how product inclusion can be scaled. Technology is not the only industry in which product and marketing teams need to get diverse perspectives to the table at critical moments in the design process. Any team that is creating a product or service needs to think about this in some capacity. The Gap team walked us through the ways they promoted diversity, equity, and inclusion in culture and product. Here are some highlights of what we learned from our partners at the Gap:

- The fashion industry is thinking about product inclusion as it relates to colors and palettes.
- Organizations have the opportunity to scale product inclusion past tech—the idea is to break down the entire end-to-end process, from ideation to launch, and find those critical inflection points where an inclusive lens needs to be prioritized.
- Senior leader buy-in is critical to move the needle with product inclusion.
- Each industry has different lead times—for the fashion industry, as they work by seasons, product inclusion needs to be intentional and prioritized up front, so it can cascade throughout the process, because physical goods are being made and numerous people and partners touch the actual end product.

Bahja focused on one inclusion initiative in particular—the Color Proud Council she formed within Gap to bring diversity into the business. The Color Proud Council's overall mission is to integrate diversity and inclusion into the entire pipeline through talent acquisition and retention and education. The Council focuses on all dimensions of diversity, including (but not limited to) gender, race, ethnicity, body type, sexual orientation, age, religion, gender, and those with disabilities (see the nearby sidebar).

The Color Proud Council

—Bahja Johnson, Director of Fashion Merchandising at Gap Inc.

The Color Proud Council started as a personal mission but grew to a company-wide movement. As a woman of color, a Black woman specifically, I grew up in an environment in which I rarely saw people who looked like me. The images celebrated on TV screens and in magazines portrayed an ideal of beauty that seemed unattainable to me and directly affected how I viewed myself. My experience with clothing also had a major impact—I wanted to look and feel good in the clothes I wore, yet it was so hard to find styles that fit me . . . styles made with me in mind. I promised myself that if I were ever able to make a difference in this respect, I would do so. The Color Proud Council is me fulfilling that promise.

I co-founded the Color Proud Council along with my best friend and fellow merchant, Monique Rollocks. We started at the company together, and along the road faced similar experiences as Black women and product leaders. Most glaringly, we found ourselves to be two of very few people of color on our respective product teams. Zooming out, we then recognized that across Gap Inc.'s family of brands, there weren't enough people who looked like us making decisions about design process and product offerings. In such an environment, we found it wasn't always easy, or comfortable, to bring our "other" to work.

I raised these concerns to Mark Breitbard, President and CEO of Banana Republic, who not only listened, but fully supported my idea to form a council. Today, the members of the council represent all of Gap Inc.'s brands and are passionate leaders across diverse backgrounds and "others" for whom this work is personal. We work together to ensure inclusivity and customer-centric decisions are at the heart of the company's work, from talent recruitment to product design and marketing.

The True Hues collection from Banana Republic, featuring an inclusive range of nude necessities (shoes, underpinnings, and intimates) for women, is a great example of how the council's influence has directly impacted the business.

This product launch took a different approach from the beginning, as we knew it had to reflect the diversity of our customer base. Starting from meeting with our in-house colorist to match the hues across various skin tones, to shooting the product on a diverse set of models, we wanted to ensure that inclusion was at the forefront of the product lifecycle.

As far as performance, True Hues knocked it out of the park. It dominated on social media, performing significantly above targets we set for key metrics (impressions, effectiveness, reach, saves, and engagement rate). It is our third most commented post of the year (as of June 2019) and the most re-posted social campaign of all time. From a financial standpoint, it also beat all targets in key product metrics (retail sales, unit sales, and gross margin) by nearly triple digits.

Inclusion within the retail industry is now table stakes, and companies that don't understand this and don't act with intention will fall behind. Customers are voting with their values as much as their wallets, and are more outspoken than ever when they feel that a company has missed the mark. That said, their "ask" isn't onerous—they just want to see themselves reflected in their clothes. I truly believe it's our job to make that happen.

What I love about this example is that expanding the hues that Gap focused on led to increased engagement and bottom line sales. By allowing people to be able to walk into a store and find their version of "nude," Gap has done well for the business and good for the customer.

Fast forward to April 2019, when Gap launched its line of inclusively designed products across the board. How exciting! Bahja posted:

> To anyone who ever struggled to find their shade of "nude":
> We got you.
>
> #diversity #inclusion #productinclusion #diversitymatters #inclusionmatters #representationmatters #design #fashion #retail #itsbanana

I was elated by this post. How many of us have gone into a store and looked for a neutral toned anything and could not find anything in our shade? I think this happens predominantly for people of color, who may have a hard time finding these colors because the nude colors or "flesh" colors are usually closer to pink or peach, but this can happen for anyone who does not match the limited tones available.

Product Inclusion at Dyne Life

Another example of the work we've partnered on in the fashion space has been a collaboration with designer Chris Bevans and his fashion line Dyne Life. For his Spring/Summer 2019 launch, instead of a traditional runway show, he showcased his line on underrepresented models through a digital mural composed of ChromeOS and Android tablets reaching all the way to the ceiling, all powered by Google Cloud.

We loved partnering with Chris because he's so highly esteemed in the fashion world (having dressed a number of celebrities including Jay-Z), and he's passionate about exploring the intersections of fashion and technology. Working with Chris was a true magic moment come to life. Googlers, led by Jamie Rosenstein Wittman, came together to bring this to life.

Exploring the intersection of tech and fashion

—Chris Bevans, Fashion Designer, Creative Director

I set out to build a technical sportswear brand using advanced materials and smart fabrics and aligning myself with manufacturers that support ethical and sustainable business practices at their core, along with being innovative in their approach to manufacturing garments. I wanted to be able to tell the story and give visibility to our customer base on our practices and partnerships.

I was introduced to Near Field Communication chips and found a way to embed them into our clothing so that we're able to interact with our base and educate at the same time, giving people the ability to learn about the garment they're interested in buying in real-time.

Partnering with Google Cloud was perfect, because it allowed us to stream our campaign video in our beautiful Paris showroom. Of the multiple brands hosted there, we stood out because of the digital mural, which displayed across about 40 tablets that synced with the cloud to create the ideal backdrop for our collection (see Figure 12.1). The tech support that the Google team showed us was fantastic, and we hope to do it again in the near future.

Figure 12.1   DYNE–GOOGLE digital campaign video.

(continued)

(*continued*)

Diversity in fashion for me is about giving everyone access to help support their vision. When you say diversity, you can talk about race, but there is also a lot of young talent that simply don't have access. I think about how Google recognized me as a leader in sportswear and gave me access. Creating space for more diversity within companies is what shapes culture and changes the world.

I use this example to illustrate that product inclusion may not always be about catching a huge mistake in your work. It can also be expanding how you think about collaboration and the use of your products. Having a wall made up of ChromeOS and Android tablets and having Google Cloud power a show at Men's Fashion Week in Paris was unexpected and expanded how the fashion industry thought about fashion and about our product suite. And we did it while partnering with, learning from, and supporting a fantastic designer from a community underrepresented in both fashion and tech. Again, asking "who else?" in terms of partnerships and support can expand your reach and enable you to show up authentically for underrepresented audiences.

Being able to work with a brand that is trusted in the underrepresented community while operating at the intersection of tech and fashion in a futuristic manner was a delight for our team. The Bevans-Google partnership is a great demonstration of how diverse perspectives can enrich products and enhance the consumer experience.

CHAPTER 13

Looking to the Future of Product Inclusion

Product inclusion is trending strong. Organizations are just beginning to wake up to the fact that customer service means serving *all* customers, regardless of age, race, ethnicity, gender, socioeconomic status, location, language, or other dimensions. They are just beginning to realize the many benefits of building *for everyone, with everyone*—increased innovation, productive new partnerships, expansion into underserved markets, positive word-of-mouth, and more. But what it really comes down to is building truly inclusive products that enrich the lives of your users.

In addition, consumers are just beginning to realize the power they have in shaping an organization's policies, processes, products, and services. They are beginning to vote with their values, their wallets, and their ratings and reviews to drive the success of organizations that are sensitive to their needs and preferences, at the expense of organizations that aren't.

The momentum in favor of diversity, equity, and inclusion is growing and is likely to accelerate as demographics shift and the competition for historically underrepresented consumers increases. In the two and a half years I've been leading product inclusion at Google, I have observed a groundswell in the pace, rigor, and enthusiasm for the work, and this trend extends beyond technology. In fashion, medicine, art, and other areas, more and more people are applying an inclusive lens to their work or discussing plans to do so.

This movement may seem surprising on its surface, but what is truly surprising is that it has taken so long. Every company wants more customers, more revenue, and higher profits. By offering products that appeal to a broader population of consumers, a company can achieve all three goals. Building for everyone, with everyone, is the right thing to do and the obvious path to future growth. It's exciting that you can "do well and do good."

In this chapter, I look at the future of product inclusion as it is unfolding in a range of industries and share other perspectives on product inclusion.

Sharing Perspectives on the Future of Product Inclusion

Remember that this isn't the work of one person, it's the work of everyone. A team mentality is essential to drive product inclusion work forward; ensuring that multiple people are committed to and held accountable for building more inclusively is critical to sustaining the effort. When deadlines, priorities, personnel changes, or vacations arise, established polices and processes must be in place to maintain forward momentum. This work succeeds most when people and processes are aligned with the objective of delivering inclusive products to consumers.

Astro Teller is captain of Moonshots at Google X—a diverse group of inventors and entrepreneurs who build and launch technologies aimed to improve the lives of billions of people. He talks about the key role diverse teams play in driving innovation:

> The lone inventor having a eureka moment is largely a myth; innovation comes from great teams where everyone feels comfortable raising questions and sharing their views. The more people a project has from a wide range of backgrounds and communities, the more fresh perspectives and creative ideas we can generate—and the better we'll all be.

Regardless of what you're creating or what industry you're in, multiple people at multiple touchpoints are involved in the product design process. Engineers, product managers, marketers, researchers, support personnel, and more all contribute to building products and delivering them to consumers. No single individual owns the process from ideation to launch. This reality poses a challenge and an opportunity for all organizations that are trying to make product design and development more inclusive. The challenge is to get everyone involved to commit to product inclusion. The opportunity is the result of the innovation born of a commitment to product inclusion.

Because no single individual owns the process, you can make a difference regardless of your position in the organization. Whether you are a member of the product or marketing team, you can contribute to product inclusion. If you are in human resources, you can work toward making the organization more diverse. If you are a member of a historically underrepresented group in any area of the organization, you can contribute by being an inclusion champion and helping create more innovative products for everyone.

Product Inclusion Across Industries

Perhaps the best way to get a glimpse into the future of product inclusion is to look at how it is currently trending across a broad range of industries. When you see the work being done in a variety of industries and organizations, you begin to see where the momentum is leading.

In the following sections, I present a number of examples that reveal the growing interest in product inclusion in industries ranging from academia and architecture to toys, fitness, entertainment, medicine, and more.

The companies I profile below are on a journey, and taking the first step with intention is critical to building more inclusive outputs. They continue to learn and develop their strategies as it relates to building inclusively.

Eyewear

Like cosmetics and fashion, eyewear is an industry that is perfectly suited to benefit from product inclusion. After all, everyone's head, face, eyes, ears, and nose are shaped differently, and everyone has a preferred look. Glasses designed to look great on one person would be a disaster on another person.

Warby Parker is a socially conscious eyeglass company. Because almost one billion people worldwide lack access to glasses, the company partners with non-profits like VisionSpring to donate a pair of glasses to someone in need for every pair it sells. The company's social consciousness extends to product inclusion; its mission is to provide great looking, high-quality glasses at affordable prices to everyone. I was invited to Warby Parker by Christina Kim, the head of one of Warby Parker's Employee Resource Groups, and spent a few hours learning about their approach to product inclusion from Christina and VP of People Chelsea Kaden (see the nearby sidebar).

Inclusive design at Warby Parker

—Chelsea Kaden, VP of People at Warby Parker

Since our co-CEOs launched Warby Parker in 2010, we've strived to make our shopping experiences as easy, convenient, and fun as possible. To do that, focusing on inclusivity across all of our platforms (both online and offline) is not just important—it's mission critical.

Glasses are the first thing someone sees when they look at you. They help define your style and are an incredibly personal product. Glasses wearers not only want their frames to feel great and function properly—they want them to look amazing, too. It's no secret that everyone's face is unique, and it's up to us to design for that. As the business has grown, we've introduced more bridge and width options (and kids products, too!) to accommodate as many customers as we can. We know we still have work to do here and look forward to continuing to broaden our assortment.

We introduced Employee Resource Groups (ERGs) pretty recently and continue to think through how ERG insights can inform projects and strategic decisions. Our resource groups are employee driven, so for now, it's quite easy and natural for ERG insights to make their way back to company-wide initiatives.

NRODA Founder and Designer Samantha Smikle has been committed to product inclusion even prior to launching the business. (NRODA is "adorn" spelled backwards.) Her luxury eyewear company has adorned some of the most familiar faces, including those of celebrities Snoop Dogg and Rick Ross. This successful brand leans into diversity and inclusion, commanding revenues in the six figures after only two years in business. According to Samantha, "People of all backgrounds, ages, and demographics take their eyewear seriously. I love that more brands making quality eyewear at affordable prices are emerging to make eyewear even more inclusive."

Appealing to a wide audience has proven successful, while providing a way for everyone to feel powerful and beautiful regardless of their race, ethnicity, gender, or other differences. Few brands think about full representation and penetration of eyewear for underrepresented users, and because of that, Samantha has been able to tap into a niche market.

Medicine/Healthcare

Healthcare providers are well aware of the challenges of diagnosing and treating a diverse population of patients. Each individual is different in terms of their physical and psychological makeup, their culture, the way they communicate, and so on. Unfortunately, the population of doctors, researchers, and research participants is not nearly as diverse as the patient population, so disconnects are common between doctors and patients, symptoms and diagnoses, diagnoses and treatments, and treatments and outcomes.

Some progress is being made as it relates to representation (60% of younger doctors are women, for example) as the need for doctors has opened the profession to people of different races and ethnicities, etc., and as more studies are being conducted to explore differences among patients. Here are a few key areas of medicine/healthcare that could benefit significantly from the application of an inclusive lens:

- **Cross-cultural medicine:** Culture and language influence how a doctor views disease and how a patient views and experiences illness.

A doctor's cultural competency and ability to bridge any language barrier plays a key role in whether a patient from a different culture receives effective treatment and adheres to the recommended treatment plan. (See the nearby sidebar.)

- **Machine learning (ML) and artificial intelligence (AI):** Computer models have the potential to help doctors increase their diagnostic accuracy, but these models must be trained on data representing the diversity of the human population. Bias in selecting the input data will create a biased computer model that is likely to be less accurate when diagnosing patients from underrepresented populations and recommending treatment protocols.

- **Personalized medicine:** Modern technology has enabled doctors to recommend treatments that are likely to be more effective based on the individual's genetic makeup. However, much of the research is based on blood and tissue samples from people of European descent. Research needs to be expanded to underrepresented populations, so they, too, can benefit from the available technology. (See the nearby sidebar.)

Considering the cultural context in medicine

—*Dr. Shawn L. Hervey-Jumper*

Cultural context can be very important because we experience life and react to events based on our upbringing and past experiences. For example, catastrophic injuries such as strokes and brain hemorrhages can leave a patient unable to make medical decisions for themselves. In this setting, healthcare teams rely on a patient's family to make treatment decisions on behalf of their loved one. Patients and families from different backgrounds rely on different strategies and resources to navigate these life-changing moments.

Recently, I treated an elderly gentleman who suffered a large brain hemorrhage after a fall. The brain injury left him unable to communicate or breathe without the assistance of a ventilator. The patient and his wife were the elders of their family and had no living children. His wife was stunned and overwhelmed by the trauma of finding her husband of over 50 years unconscious on the bathroom floor.

When they arrived at the hospital, our medical team had to relay facts about the severity of his injury and explain treatment options to his wife, after which we needed to seek direction on how the patient would wish to be treated.

As a surgeon, I know the published literature, how I would like to be treated, and how I would discuss care in this setting with my own family, but in this case the cultural context was different. I was fortunate to have a member of my team who understood the couple's culture and felt empowered to share insights with me and the rest of the team. As a result, my ability to help guide this family through the numerous complex decisions required for his care was greatly enhanced. I believe that this patient and his family received more effective and compassionate care because of the diverse perspectives of our team.

Delivering personalized care to diverse populations

—*Dr. Shawn L. Hervey-Jumper*

Advances in biomedical research have and will continue to shape treatment options for patients. This is particularly true with respect to *personalized medicine*. In many instances, healthcare providers offer treatments influenced by the genetic make-up of an individual, especially in the field of cancer. An increasing number of cancer drugs are directed toward an altered pathway specific to an individual's genetic make-up and tumor genetics. This is the truly exciting part of biomedical research and explains why select types of lung cancer and melanoma have gone from being terminal diagnoses to virtually cured. Patients are therefore counseled about prognosis

(continued)

(continued)

and receive treatment recommendations based on large genetic epidemiology studies.

However, all too often our best studies are based on data that doesn't mirror the population we serve. Biomedical research in the United States, including clinical trials, is overwhelming based on enrolled patients of European descent. Clinicians are therefore forced to make inferences for minority patients based on best available data, knowing that we may be wrong.

For example, I am a neurosurgical neuro-oncologist who specializes in the removal of brain tumors. A tumor called glioma is the most common tumor I treat, and over the past 10 years our field has learned that patients with tumors which look the same *histologically* (under a microscope), actually have very different diseases and responses to treatment based on genetic subtypes. These effects are so important that the way we grade tumors now reflects these differences. However, the truth of the matter is that the landmark studies we use to prognosticate and guide treatment decisions are based on brain tumor and blood samples from individuals overwhelmingly of European descent.

This point matters because when I counsel a patient who was not represented in the study population I'm making assumptions. Will survival and treatment response rates for our patients from China fall in line with published results? How might results differ if comparing published results with patients from communities in India and Sub-Saharan Africa?

I mention this not to cast doubt on these very elegant results and brilliant studies, but it's critical that we appreciate the diverse make-up of society. Companies will put enormous resources into drugs based on available data that might not ultimately be (as) effective for all patients. Therefore, having decision-making teams which are diverse with respect to gender, ethnicity, social, and economic background may lead to better treatments for patients, which is the ultimate goal.

Architecture

A senior leader at Google once mentioned to me that a very low percentage of architects were women, and even fewer were women of color. In architecture, diversity may not seem like a big deal until you begin to consider the various groups that can be marginalized by architectural designs—particularly people with disabilities, those who are significantly larger or smaller (or taller or shorter) than average, among other demographics.

Have you ever tried to open a door that was too heavy or tried to reach something in a cabinet that was too high? Have you been in a dressing room with a stall door or walls that were not tall enough to ensure your privacy? Have you had to tilt your head or crouch down when walking through doorways to keep from hitting your head? People of "average" build do not have these experiences, but for many people, these are daily inconveniences that often make them feel marginalized.

I recently visited the Stuart Weitzman School of design at my alma mater, the University of Pennsylvania, and was heartened to hear students wanting to continue the conversation around product inclusion and building inclusive spaces for everyone. The next generation of builders hopefully will think proactively about how spaces affect different people and the deep opportunity and responsibility they have to make that experience as equitable and inclusive as possible.

Fitness

During a recent visit to a local gym, I noticed that the instructional drawings included very few women in general and no people of color. What message does this send? Are trim, men the only people who work out? I knew this could not be true because looking around me I could see that at least half the people working out at this gym were women, and they represented many different ages, shapes, and sizes.

How great would it be to see more inclusive depictions of people who could benefit from these products? The world is full of athletes of all colors, sizes, genders, abilities, and shapes, so making the industry more inclusive would not require a huge effort—just a simple shift in focus.

I've been fortunate enough to be introduced to Mary Keane Arenson, lululemon maven and community disruptor, who leads with an intentional focus on creating for intersectional and diverse groups of people in the San Francisco community.

lululemon's commitment to improving equity and learning from past challenges is felt, both from the work they do to foster community, and in key process like inclusive marketing and design. While no company is perfect, a dedication and focus to bring more voices to the table is what leads to lasting change.

Mary's (and the rest of her team's) focus on establishing authentic relationships enables the company to build connections within communities across North America. Across multiple dimensions, the luminaries program shows up in a real way, and in turn breeds brand trust with members of the luminary community.

According to Mary, "I'm fortunate to work with and for lululemon; a company that invests in igniting communities through opportunities to sweat, grow, and connect—together. Together, lululemon and our communities have the opportunity to elevate the world by unleashing the full potential within every one of us."

The focus on building community with everyone is not only the right thing to do, but also makes business sense. The stock has rallied almost 70 percent in 2019, and has loyal followers across gender, geography, and more.

Toys

Toys, dolls especially, can be integral in shaping children's vision of who they are and what is possible. Not seeing themselves reflected in toys can negatively impact children's self-esteem. Conversely, seeing themselves reflected in their toys nurtures feelings of belonging and builds confidence.

Amy Jandrisevits believes that all children should be celebrated and be able to look into the face of a doll and see themselves. In fact, she made that her mission by creating custom dolls she calls A Doll Like Me (see Figure 13.1 and the nearby sidebar).

Figure 13.1   Keagan enjoys a doll from A Doll Like Me.

Giving all children a chance to see themselves in the world

—Amy Jandrisevits, Founder of A Doll Like Me

One of the prerequisites of play therapy is for the child to be able to see him/herself in the toys s/he plays with. When I did pediatric oncology social work, it was even more important since children with medical conditions are underrepresented in the toy market.

(continued)

(*continued*)

When you look at the "whole health" model of a child, you have to include mental and emotional health, which are integral to physical well-being. The power in seeing yourself represented and depicted in major channels is underappreciated.

Dolls provide opportunities for physical comfort, personal validation, and play therapy. My master's thesis was about the healing power of play, and dolls are such an integral part of that; unlike most toys, they are a human likeness. Children should look into the face of the doll that they are playing with and be able to see their own. For that reason, it is my heartfelt belief that dolls should look like their owners and dolls should be available in all colors, genders, and body types.

We do kids such a disservice in not offering a wide variety of toys. In an ideal world, limb difference, body type, medical condition, birthmarks, and hand differences would be as accepted as all of the other things that make us unique.

The recent media exposure has brought this discussion to the table—*WHO* do we see and *HOW* do we see them? While having a look-alike doll can be so incredibly validating for a child, seeing them together also helps normalize this on a bigger scale. We have to be able to look at these kids and fall in love with those sweet faces to actively participate in the paradigm shift. We won't fight for the things we don't connect with.

For many of these kids, the positive public response has been so incredibly validating. For the first time, the conversation reads: "the adorable little boy with the doll" and not "what's wrong with him?" That is a wildly different narrative. The media exposure is changing the narrative for so many kids, and I am thrilled that dolls have been a part of the discussion.

The toy industry represents $20 billion in the U.S. alone; as demographics shift, ensuring that all children are reflected in the toys they play with will become increasingly important. Amy has a multiyear waiting list and has been featured in *O* magazine. People yearn to be seen, no matter what age they are, and Amy's dolls are paving the way for many to see themselves reflected in toys.

Movies and television

Diversity is increasing in the movies and television in regard to casting, but more needs to be done. The people writing, producing, and directing in the U.S. still don't reflect the diversity of the general population. As a result, you may see more characters of different races and nationalities, but the stories do not reflect a great diversity of cultures, experiences, and perspectives.

Notable exceptions are movies such as *Black Panther* and *Crazy Rich Asians*, which transcend race, age, geography, and more as it relates to connecting with audiences. General audiences are attracted and excited to see new characters on screen and to gain insight into the unique experiences and perspectives coming out of different cultures. These movies also help to make the business case for inclusion. *Black Panther* claimed the second-biggest four-day opener of all time at the domestic box office with $242.1 million, just behind *The Force Awakens* ($288 million).[1] *Crazy Rich Asians* grossed $174.5 million in the U.S. and Canada and $64 million elsewhere for a worldwide gross of $238.5 million, against a production budget of just $30 million.[2]

Crazy Rich Asians was the first Hollywood movie in decades to have an all-Asian cast. The success of both movies shows that casting people of color should not be the struggle that it is. But it's more than just having a diverse cast that made these movies successful. Having writers and directors of color added to the authenticity and cultural richness of both pictures, so it's no shock that these movies did as well as they did. But it's important to note that though these movies centered on people of color, people who don't identify as a person of color also were deeply drawn and connected to the narrative.

I caught up with Vivaciously Vivian (Vivian Nweze) a young media tour de force, and spoke with her about how she sees the entertainment industry shifting and the value of having diverse representation

[1] https://deadline.com/2018/02/black-panther-thursday-night-preview-box-office-1202291093/.
[2] https://en.wikipedia.org/wiki/Crazy_Rich_Asians_ https://www.boxofficemojo.com/release/rl1157858817/

throughout the production process. She was very upbeat about the future of inclusion in the entertainment industry, as expressed in the nearby sidebar.

Increasing diversity in the entertainment industry

—*Vivian Nweze, TV Host, Producer, and Entertainment Journalist*

I think there is a lot of work to be done to make the industry more inclusive, but right now it's a great time to be a person of color in entertainment. We're starting to see more seats at the table, aside from having "the one designated Black person." Last week on E! News, Nina Parker and Zuri Hall co-anchored the show. It was the first time two Black women anchored a mainstream entertainment news show! More work needs to be done on screen to reflect the various shades of people of color. Every young Black female lead doesn't have to be mixed or light skinned. Now that we're getting more opportunities, we need to make sure that we're showcasing all of our skin tones.

I wanted to be a part of the entertainment industry because we are the keepers and shapers of culture. Diversity is absolutely necessary to do that. We have to reflect the world we live in.

When I was at *Access Hollywood* we did a story covering Cardi B right before she fully blew up. Before it aired I saw it and told the executive producer that it didn't capture the hype building around her. There were enough young people and people of color in the office to reference when dealing with artists our older producers knew nothing about. Next thing you know, I'm being interviewed for culture commentary about her rise to fame. There were numerous times at *Access* where many senior producers wouldn't feature or consider high profile celebrities of color because we lacked representation at the higher levels. When I became a producer and started working on-camera, I became a voice to champion the diverse entertainers who were actually leading entertainment as the creators of what's "cool."

Bringing an international lens to TV screens

—*Michael Armstrong, Executive Vice President, Paramount*

Designing a new television channel for a global audience means being deliberate about how the brand will be perceived in every country where it launches. Thinking about geography was top of mind for our team.

With Paramount Channel we began by creating a look and feel that would be consistent no matter if you were in a rural European village, a Latin American city, or an Asian metropolis. Knowing that on-screen text graphics would appear in Hangul, Chinese, Latin, Cyrillic, Greek, Arabic, and a range of other written scripts, I partnered with our internal World Design Studio to select an easily legible and very accessible font (DIN). This allowed our local channel managers a consistent look when using their local script without sacrificing the uniformity we desired for the brand.

Next, we created a set of channel indents for on- and off-air promotion of the network content. We chose to create animated, genre-specific scenes in silhouette rather than casting actors. By using silhouettes, we were able to more broadly represent the ethnic and racial makeup around the globe. We aligned across the globe to use so that we were not unintentionally excluding, while allowing teams to create new assets that further reflected their local population, instead it made certain that we were not unconsciously creating exclusionary assets for any one of our regions.

Expanding the Black Entertainment Television (BET) brand around the world created a very different challenge. When thinking about our international strategy, we knew that there were people of color (particularly Black users) who had limited options of seeing themselves, reflected in popular media. But having content featuring African-American talent was not a panacea. While eventually developing shows with local talent in markets such as France, Africa, and the United Kingdom, we needed a cost-effective solution that

(continued)

(*continued*)

created an authentic connection with the new BET channels and its viewers.

Our first channel rollout happened in the spring of 2008 in the UK. Ahead of the launch, we began by redesigning our show promotional spots on the content we exported from the U.S. We started by hiring local British voice talent whose pronunciations and voice inflection would ring familiar to our targeted viewers. Talent was given the green light to change the scripts for both on-air and radio spot copy because they knew best how to convey a message in their home market. We also changed the on-screen text graphics to reflect the differing date and time formats on all on-air and off-channel copy.

The most important focus was on our interstitials; the short-form content that runs often between our long-form shows. This gave us an opportunity to further bring local faces and voices to our screens. From local music artist features, adding international categories to our award shows, and celebrating local entrepreneurs and influencers, we used every possible moment on our channels to create a cultural tapestry that reflected underrepresented users in an authentic way.

The Paramount channels are seen in more than 120 countries and territories via 16 locally programmed and operated channels, reaching more than 160 million homes outside the U.S.

BET International reaches 45 million households via 3 locally programmed and operated channels in 64 territories worldwide.

Academia

Ontario College of Art and Design (OCAD) is Canada's largest design university. I spoke with Dean of Design, Dori Tunstall, to understand the role that higher education can play in driving the future of product inclusion. Dori highlighted OCAD's product inclusion principles, which include the following:

- **Graphic design:** Minimizing harm and emphasizing the good we can do. Offering a strong focus on design process, typography, image-making, and critical, strategic thinking. Mindful design systems for everyone and everything.
- **Illustration:** Developing one's individual artistic expressions to critically reflect on society. Centered on knowledge and skills that create effective, communicative, and artful images that function as a complement to the written word.
- **Social innovation design pathway:** The social innovation design pathway offers students a series of courses that connects their home discipline and studio work to the broader context of social action and purpose in design.

The intentionality behind the principles for each of these areas shows clear and intentional focus of ethnicity, accountability, and human-centered design. By naming and centering equity in their mission, vision, and courses, they will likely usher in many of the product inclusion revolutionaries who are needed to shape the future of industries around product inclusion.

Fashion

Chapter 12 highlights fashion and retail as important industries where product inclusion will have a significant positive impact, but I have more to say about the future of fashion. I have talked with several thought leaders in the fashion world about the future of product inclusion in their industry. In this section, I share their insights.

Two years ago, Jackson Georges and I ran the Griots at Google event with the Council of Fashion Designers of America (CFDA) and it was there I met Chrissy Rutherford, Special Projects Director of Talent and Social at Bazzar.com, the e-magazine version of *Harper's Bazaar*. Chrissy is inspirational, intentional, and passionate about bringing underrepresented voices to the center of her and her team's work. Chrissy shared how she leverages an influential platform to level the playing field for historically underrepresented talent in the fashion world for the benefit of readers (see the nearby sidebar).

Shaping perspectives through fashion news and commentary

—*Chrissy Rutherford, Special Projects Director of Talent and Social at* Bazaar.com

As the Special Projects Director of Talent and Social at Bazaar.com, part of my job is staying current on what's cool, who people are talking about, who's setting trends—and so often Black people are behind viral trends or become overnight pop culture phenoms, but they don't always get the credit or they get lost in a sea of appropriation. I want to make sure that Black women get the props they deserve, and/or their stories are being heard. I'm always suggesting talent for video or photo features, so I'm mindful of representation and inclusion on our site. I'm also just always on the lookout for interesting stories.

I truly believe that the fashion industry has the power to change the way marginalized groups are seen and valued. This industry boasts so many marketing masterminds—from magazine editors to advertising executives. We can convince the masses to buy the latest It Bag, or that pistachio green is the color you need to be wearing; therefore, we also have the ability to shape how we view Black women, curvy women, the trans community, and more.

Luckily, I work online, so we always have hard evidence for the success of our stories—page views! For example, Rachel Cargle's first essay performed almost 10 times better than our average views on a single piece of content. So we have definitely proved that Black women's stories matter, and that we have an audience for them.

Another example of product inclusion in the fashion industry is the push to get more designers of color visibility in order to have more access to consumers and more mainstream fashion. Brandice Daniel founded Harlem's Fashion Row, a company that has prioritized bringing diverse voices to the forefront and ensuring that fashion is representative of multicultural designers. She's worked with LeBron James, Nike, and many more, and has a passion for bringing underrepresented designers to the forefront. In the nearby sidebar, she shares a success story of partnering with Nike, LeBron James, Melanie Auguste, and three designers from underrepresented groups to build a new shoe that sold out in a matter of minutes!

Increasing the diversity of fashion designers

—Brandice Daniel, Founder, Harlem's Fashion Row

In 2018, we got a call from the brand manager of the LeBron James brand at Nike. Melanie Auguste is one of the few AfricanAmerican women to have held that position, and it was because of her background that she saw an opportunity in a very organic statement that LeBron James made that Black women are some of the strongest people on earth.

When Melanie heard the statement, she saw an opportunity and called us to discuss a partnership. I selected three designers for Nike to choose from and they chose all three. We went to Portland, and the team just clicked. Once we all told our stories, we were bonded and felt as if we were working on something much bigger than a shoe.

LeBron James launched the shoe at our NYFW event in 2018. The shoe went on sale a few days later and sold out in less than five minutes. Black women were eager for a brand as major as Nike to speak to them in a way that was real. There was no better way to do this than to have three African-American women who've had to overcome adversity, lack of resources, and tragedies to do this.

It's interesting because in the last two years, consumers have been asking me more and more about supporting designers of color. In 2019, designers of color still represent less than five percent of designers in retail headquarters, and it's less for luxury brands. It seems that now, more than ever, consumers are very intentional about what they buy and who's behind it. The reaction has been amazing.

Yet another example of bringing an inclusive lens to the fashion industry is Claire Sulmers, who created the mega successful *Fashion Bomb Daily*—a platform that showcases the work of multicultural influencers, celebrities, and other fashion notables. Launched in August 2006, *Fashion Bomb Daily* provides daily doses of chic to savvy fashionforward subscribers.

Recognizing there were few print and online sources for multicultural fashion, Claire leveraged her interests in style and writing to create the number one online destination for global chic seekers with a penchant for all things fabulous. Through her blog, Claire celebrates different facets of what fashion, luxury, and style can look like.

Starting my own fashion magazine

—*Claire Sulmers, Founder, Fashion Bomb Daily*

When I started *Fashion Bomb Daily* in 2006, seeing Black women in high fashion editorials or on the cover of fashion magazines was a rarity. I was always a fashion girl and never saw women like me being catered to, highlighted, or celebrated. So I decided to start a blog that represented the change I wanted to see in the fashion industry.

Fashion Bomb Daily came about because although I had graduated from an Ivy League school and done all the right internships, I still could not even be considered for a job at a fashion magazine. *Fashion Bomb Daily* was created as a hobby to show my writing ability—sort of as a resume builder or an extra credit project for editors to see what I was capable of. It started as a blog and became a business thanks to the digital revolution.

I think the fashion industry operates on an archaic prism of beauty. Now our purchasing power has increased, but the global fashion industry will sometimes not even cast a cursory glance towards women who are brown or different sizes or have various backgrounds.

Also, several missteps by large companies show an ignorance towards how culture is reflected through style (and how to be respectful of that relationship). Inclusion is important because we are here, we are consumers, and our dollars and our voices matter.

I now have 1.3 million Instagram followers, millions of impressions, recognition from *Ebony*, *Essence*, and *Teen Vogue*, and scores of awards and acknowledgments. This proves the business case for inclusion in the fashion industry.

Claire has created a conversation, a platform that celebrates diversity across multiple dimensions, including race, ethnicity, and size. Millions of people look to her to see themselves reflected, and her influence spans the globe.

I could continue with examples from more industries, but in every case, the moral of the story is always the same—doing what's right for consumers is what's best for organizations. The cost of integrating product inclusion into what an organization does is far exceeded by the benefits in innovation, revenue, growth, positive word of mouth, and more!

I spoke to Cara Shortsleeve, a former Google director I deeply admire, about the future of product inclusion. Cara is currently CEO of The Leadership Consortium (TLC). Conceived by renowned Professor Frances Frei, TLC is a leadership accelerator for companies who embrace critical talent and consider diversity and inclusion to be a lever for improved performance. Cara shares her insights about TLC, and the future of product inclusion in the nearby sidebar.

A future with and without product inclusion

—Cara Shortsleeve, CEO of The Leadership Consortium

When asked what I think about the future of business without product inclusion, I imagine a world that is sub-optimized, a sliver of its potential self.

(continued)

(*continued*)

When I think about a business that *doesn't* embrace diversity and inclusion as a lever for performance, there is certainly a scary version of the future—in which there are dire negative consequences. But the more likely scenario in my mind is that the brilliant future of business would simply be unrealized. If we don't actively seek out and include diverse viewpoints, critical products and services won't make it to market; the products and services which do make it to market won't be as effective as they should be; companies won't be as successful as they could be; and individuals will feel less joy as they fail to realize their own full potential and the full potential of those around them.

So as an optimist, I prefer to envision the future of business when inclusive product design has been prioritized: companies with inclusive DNA will be powered by more effective teams, which will ship more successful products and services, and ultimately deliver better business results. So let's work together to make *this* version of the future a reality.

Personally, what I love most about our TLC Leaders Program is that while all participant profiles are welcome, we have a stated focus on *disproportionately* serving populations that have been *underserved* in the past—and who are vital to the future of competitive organizations. Our unapologetic goal is to accelerate this pool of critical and underrepresented leaders into bigger, bolder, and more visible leadership roles—so that our client companies see improved business results, employees within and outside our client companies see an invigorating model for success, and ultimately other businesses follow suit towards a future of dramatically more product inclusion.

Although Cara does not mention the three P's framework mentioned earlier in the book—process, people, and product—she certainly alludes to them. Inclusive *processes* that involve a more representative sample of *people* build and ship more inclusive *products*. All three P's are inextricably linked, forming a synergy that serves more customers better and drives innovation and growth.

As a thought leader on product inclusion, you may have to get creative and find ways to influence people or give them concrete recommendations on how to integrate product inclusion with their work. If you make people excited about inclusion and get them to understand that better innovations will emerge when they dream big and take action, change will occur.

So What Does All This Mean?

The people I've highlighted come from all backgrounds and walks of life. Sure, there are many that are brilliant leaders and innovators in this book who happen to be from an underrepresented background. But there are also several people who don't identify as underrepresented.

What's interesting about product inclusion is that many people who may not have seen where they fit in to the traditional diversity, equity and inclusion conversation are energized by this work. It is concrete and actionable, and they can take a few steps to change how they do their work. They can also get behind the why-they understand the end goal and the reasoning both from a business and a user perspective.

So what have we learned over the past two years of experimentation, iteration, trial and error, and a few breakthroughs?

- That diverse perspectives lead to better results and increased innovation. For all users, not just underrepresented ones.
- That ideation, user research user testing and marketing are core areas to focus on bringing an inclusive lens to your process, even if it's one commitment from each area.
- That product inclusion should be embedded into current processes, not a standalone idea or process that is tacked on at the end.
- That bringing an inclusive lens does not necessarily mean building more slowly. It's about building more intentionally.
- That billions of users are yearning to be seen in products and have the purchasing power to act when they've been brought into the fold.
- That the further away from the "default" user your team has in your mind, the more likely you are to provide an alienating or bias experience for users not like that target user if you aren't intentional about bringing their stories, needs, and core challenges into the fold at key steps along the way.

And arguably most importantly, that a business can do well and do good. Building products and services that reach more users will undoubtedly grow your business. Taking a critical look at your current process and identifying where you've inadvertently not brought in perspectives that are unlike your network will help expand who your target users could and should be.

Remember:

During ideation: Bring together a diverse set of participants to think through and intentionally expand who your target user is and what use cases there are for your product. When you are building for moms, do you mean parents, caretakers, and more? When you think about design, have you brought in multiple genders to think through priorities in functionality and design?

During user research: Have researchers from a diverse set of backgrounds performing the research and during the setup of the research, ensure that inclusion is a key facet by creating principles around how the team will work. Set goals for the participants who will provide the data that will help your product managers create the next steps for the product. If you cannot have a diverse set of researchers, think through how to expand your network (public venues, online surveys, social media, etc.) where you can get perspectives of users that can fill in gaps or areas you may not have thought of.

During user testing: Find ways to bring in diverse group of users to try out your product and provide feedback. Even if there isn't a huge aha moment that shows a large issue as it relates to one of the product inclusion dimensions, having diverse perspectives leads to diverse ideas and opportunities for you to think through new ideas. This should be a mantra on your team: different perspectives lead to better results which lead to larger opportunities.

During marketing: Tell real stories about user's lives and how your product can help make their experiences better. Tell stories that encompass multiple dimensions of diversity, because people empathize with people that they can connect to and feel and look like them. When we look at how demographics are shifting and more and more multiracial consumers are coming online and are increasing in purchasing power, it makes sense to ensure that you are connecting with those users and telling their stories with them. It also can be a costeffective way to do marketing to take real, diverse users that already love your product. This resonates with people because of the palpable authenticity.

And lastly, listen and be humble. We're all on a journey to empathize, solve problems, and tap into opportunity by doing so. Put yourself in the shoes of your users. Ask questions. And keep trying! The opportunity to build user-centric products and build better business awaits!

ACKNOWLEDGMENTS

It took a village to write this book and to create the framework that has now become product inclusion—to begin the journey, to keep it moving forward, and to transform first drafts into a complete and polished product.

Christen Thompson, the publisher who initially reached out to create this book and believed and was excited in what an introvert like me had to say about product inclusion in her own way. Thanks for being a champion and a cheerleader. To Richard Narramore, for helping kick off this project, and to Victoria Anllo for keeping us all in line and for always being so kind. To Mike Campbell, Vicki Adang, Mike Isralewitz, Koushika Ramesh, and the whole Wiley team—thank you.

Chris Genteel, who believed in this work, and believed in me. Thank you for your leadership, 20 percent projects, and counsel for so many years. Thank you for your tireless commitment to underserved communities, your decade long vision around business and the digital divide, and of course, for caring so much about all of your team members.

To Naomi Kraus from Google's UX Studio: You are like my editing fairy godmother! Thank you so much for volunteering to help get this to the finish line and for your support and reinforcement that this was important work, even when writing got tough. I'm deeply indebted to you and so very grateful.

Chanelle Hardy, thanks for stepping in and so thoughtfully ensuring that the tone and content came through authentically. I appreciate you so very much!

To Joe Kraynak, this would not be the book it is without your editing prowess—I am so lucky to have been connected to you and to have your guidance and expertise.

Of course, to my extended family who always allow me to be elevated and seen in my best light, and to the people around the world who yearn for that very feeling.

Angelo Carino and Donald Lee Bullock Jr., thanks for being amazing friends. Special thanks to Stephanie Richards and Cristina Rodrigues, my "lawyers," for your counsel. Cristina, I'm so fortunate to have a generous friend like you who will proactively volunteer to edit! Shavonne, thanks for all you do and all you did for this book and for your constant support. To Jim, Ian, and Hilary—thanks for making SF an adventure (even from afar!), and for having my back through this process and always! To bailey carrol, thanks for being a great friend and confidante. Kamil—thanks for being such an amazing cousin and friend.

Thank you to John Maeda for writing the foreword and also for answering a cold call and helping guide this journey. You will forever be an inspiration.

Thanks to my Google family, especially:

- Exec sponsors of the research: Erin Teague, Andy Berndt, Sherice Torres, David Graff, Shimrit Ben-Yair, Kat Holmes, Sowmya Subramanian.
- Thanks to the Product Inclusion team—our phenomenal 20 percenters—and to our supportive Googlers, leaders, and mentors, especially:
  - Sydney Coleman, Guillermo Kalen, Connie Chu—you two are phenomenal and are a true extension of the team. Thank you for giving so much to the work. I owe you!
  - Ty Sheppard and Lynette Barksdale for helping me think through this from the beginning, and for always giving me sage advice.
  - Amanda Kuehn—thank you for your feedback and proactive teamwork! Carly MacLeod, thanks for the support and advice!
  - To Sanjay Batra and team, thanks for being incredible partners.

Michelle Banks, Kr Liu, Manprit Brar, Amber Ibarra, Jeffrey Dunn, Anette Sjorgen, Ashley Wilson, Tom White, Jason Randolph, John Pallett, Reena Jana, Lucy Pinto, Ava Donaldson, Brandon Asberry, Salvador Maldonado, Miles Johnson, Holden Rogers, Meghan Chisholm, Parker, Dan Friedland, Alana Johnson Beale, Sadasia

Mcuttcheon, Britt Deyan; and the co-collaborators on the early integrating inclusion work: Allison Bernstein, Allison Municchiello and Randy Reyes. Guillermo Kalen, Danielle Hurhula, Heather Cain, and last but certainly not least, Victor Scotti.

Thanks to Susana Zialcito for guidance on the building for everyone vision, Natalie for your data and patience, Connie Chu for your brilliance and insight, Thomas Bornheim for your willingness to learn and teach, and the PI-UX team—this work would be nothing without you. Thanks to Lauren Thomas Ewing for mentorship and support. Seth, Michael, and Andy, thanks for always being there. To the Speechless team, thank you for your help in finding my voice authentically.

Taylor Nguyen, thanks for being such a collaborative and supportive partner and leader and helping get the word out about this work early and often!

Thanks to Nina Stille and Tomas Flier for their support, energy, and their vision; to Rags Williamson and Jon Crawford for bringing these stories to life via video; and to Alyson Palmer, for always being open to providing sage guidance.

Thanks to thought partners, leaders, and sponsors: Melonie Parker, Jeff Whipps, Corey DuBrowa, Kat Holmes, Kyle Ewing, Mona Gohil, Laura Hughill, France Olajide, Sarah Saska, all the working groups, the inclusive marketing consultants, and every team that has let us in.

Special thanks to Bradley Horowitz, Parisa Tabriz, Seang Chau, Erik Kay, Asim Husain, and Matt Waddell for all of your support!

Deep gratitude to Hiroshi Lockheimer—this work has grown so much with your support and vision.

Thank you to all of the contributors to this book—thanks for sharing your wisdom via sidebars and interviews and showing no matter what your product or service, if you have a user, you should be bringing more voices to the table.

Thanks to Jackson Georges for your incredible energy and constant support, to Jessica Mason, Jenn Kaiser, and Flavia Sekles for your patience, collaboration, and belief in this work. Jennifer Rodstrom, thanks for your incredible partnership and belief that these stories should be told. Hannah Hunt, thank you for being such an ally, friend, and champion. Mary Streetzel, thank you for your constant support, laughs and guidance.

Rebecca Sills and Amanda Keuhn—this book would not have crossed the finish line without you. Thank you for your generosity, patience, and editing wizardry.

Googlers across the company who make it happen (and more than are listed here!): Amanda Gorney, Elias Moradi, Chris Patnoe, KR Liu, Elise Roy, the PI-UX working group, Giles Harrison-Konwill, Jen Kozenski Devins, Steph Boudreau and Dimitri Proano, Yolanda Mangolini, Sadasia Mccutchen, Ruha Devanesan, N'mah Yilla, Paul Nicholas, Laura Palmaro-Allen, Anastasios Kolonelos, the product inclusion community, Mariko Cates, all of our working groups, the ERG-PI leads—thank you! Thank you to all of the leadership who allowed me to share my opinions about product inclusion and development from my experiences at the company.

Thanks to Rachel Lambert and Karen Sumberg who have navigated through so many ups and downs with me over the past 3 years—safe to say I'm not sure where I would be without you. To Anna Davda, thanks for being an incredible mentor.

Thanks to Eve Andersson for your support, guidance, and leadership, always. Leslie Leland, thanks for being a role model and fierce advocate!

Thanks to Erica Dumas, Henry Cunningham, Shante Bacon and Saptosa Foster; and to the essence-capturing dream team of Vadee Chhun, Toni Neal, and especially Shamayim. Thank you to Sarah Saska, Nanjappa Palekanda, Wieland Holfelder, Daniel Navarro, Jason Scott, and Feminuity.

Special thanks to my outside-of-Google support systems (the village it takes to help someone find their light and purpose), my entire extended family, especially Tatie Nelly and Uncle Rudolf, Tatie Marlyne, Fred Stephens, Merrin White, Kendall Brown, Mike Giannone, Katie Boyd, Danielle Hurhula, my grandparents (including the angels), and my Penn family, my teachers at Milton and Penn, Conor, and Bradford. Special thanks to Justin Reilly, who has pushed me and grown with me. Thanks to the Dartmouth, Ma crew that constantly is open to learning and growing-you know who you are.

To my Google women past and present—Whitney Moskowitz, Shoutout to the equity army-grateful for a community of learners and innovators. Heather Cain, Dominique Mungin, Suezette Yasmin Robotham, Delecia Krevet, Danielle Hurhula, Erika Bennett,

Erin Teague, Camille Stewart, Mecca Williams, Jen Sutton, Tiffany Snowden, Natasha Aarons, Erika Munro Kennerly, Suezette Robotham, and Danielle Lucq, my Odyssey crew, Lena McAfee—thanks for holding me down always. I love y'all! Thanks for guiding me through so much and for being mentors.

And again, to my family, especially Mum, Daddy, Allan, Todd, and my cousins, especially the next generation-inspired by you and excited for all that will ufold. Herc: nothing about me happens without you. I love you.

ABOUT THE AUTHOR

Annie Jean-Baptiste is the Global Head of Product Inclusion at Google. She founded the framework for product inclusion two years ago, and has helped it grow and scale both inside of Google and across various industries. She is passionate about making the web and products work for underserved communities while ensuring that Google is a place where everyone shines for their differences. Formerly, she led programs related to diversity talent management and career development within several of Google's technical product areas.

Annie graduated from the University of Pennsylvania in 2010 with degrees in International Relations and Political Science. She currently serves as an intrapreneur in residence at the University of Pennsylvania's Graduate School of Education and is a member of the IEEE's Ethically Aligned Design committee.

Annie has been covered in *Vogue, Teen Vogue, Cheddar, Digital Trends,* ABC, CNBC, *Essence,* the *Huffington Post, The Root,* the Council of Fashion Designers of America Annual Report, the *Miami Times,* the *Boston Globe,* and *Fortune* magazine.

She currently lives in San Francisco with her husband, Todd, and her dog, Hercules. She is 31 years old.

Want to get in touch? Annie's website is Anniejeanbaptiste.com or you can find her @ Its_Me_AJB on Instagram and Twitter, and let her know what you think!

INDEX